GENETICS - RESEARCH AND ISSUES

ANIMAL GENOMES UNDER THE FOCUS OF CYTOGENETICS

GENETICS - RESEARCH AND ISSUES

Additional books in this series can be found on Nova's website
under the Series tab.

Additional E-books in this series can be found on Nova's website
under the E-books tab.

GENETICS - RESEARCH AND ISSUES

ANIMAL GENOMES UNDER THE FOCUS OF CYTOGENETICS

CESAR MARTINS,
DIOGO CAVALCANTI CABRAL-DE-MELLO,
GUILHERME TARGINO VALENTE,
JULIANA MAZZUCHELLI,
SÁRAH GOMES DE OLIVEIRA
AND
DANILLO PINHAL

Nova Science Publishers, Inc.
New York

Copyright © 2011 by Nova Science Publishers, Inc.

All rights reserved. No part of this book may be reproduced, stored in a retrieval system or transmitted in any form or by any means: electronic, electrostatic, magnetic, tape, mechanical photocopying, recording or otherwise without the written permission of the Publisher.

For permission to use material from this book please contact us:
Telephone 631-231-7269; Fax 631-231-8175
Web Site: http://www.novapublishers.com

NOTICE TO THE READER

The Publisher has taken reasonable care in the preparation of this book, but makes no expressed or implied warranty of any kind and assumes no responsibility for any errors or omissions. No liability is assumed for incidental or consequential damages in connection with or arising out of information contained in this book. The Publisher shall not be liable for any special, consequential, or exemplary damages resulting, in whole or in part, from the readers' use of, or reliance upon, this material. Any parts of this book based on government reports are so indicated and copyright is claimed for those parts to the extent applicable to compilations of such works.

Independent verification should be sought for any data, advice or recommendations contained in this book. In addition, no responsibility is assumed by the publisher for any injury and/or damage to persons or property arising from any methods, products, instructions, ideas or otherwise contained in this publication.

This publication is designed to provide accurate and authoritative information with regard to the subject matter covered herein. It is sold with the clear understanding that the Publisher is not engaged in rendering legal or any other professional services. If legal or any other expert assistance is required, the services of a competent person should be sought. FROM A DECLARATION OF PARTICIPANTS JOINTLY ADOPTED BY A COMMITTEE OF THE AMERICAN BAR ASSOCIATION AND A COMMITTEE OF PUBLISHERS.

Additional color graphics may be available in the e-book version of this book.

LIBRARY OF CONGRESS CATALOGING-IN-PUBLICATION DATA

Animal genomes under the focus of cytogenetics / Cesar Martins ... [et al.].
 p. ; cm.
 Includes bibliographical references and index.
 ISBN 978-1-61209-390-1 (hardcover)
 1. Cytogenetics. 2. Genomics. I. Martins, Cesar.
 [DNLM: 1. Animals. 2. Chromosomes--genetics. 3. Chromosome Mapping. 4. Cytogenetics--methods. 5. Genome--genetics. QU 470]
 QH441.5.A55 2011
 572.8'61--dc22
                      2010049667

CONTENTS

Chapter 1		Chromosomes and Genomics	**1**
	1.1.	The Chromosome History	1
	1.2.	Basic Features of Chromatin and Chromosome Structure	2
	1.3.	A Brief History of Karyotypes under the Focus of Basic Cytogenetics	8
Chapter 2		In the Animal Genomes	**21**
	2.1.	Chromosome Records in Animals	22
	2.2.	Genome Features	23
	2.3.	Driving Forces upon Genome Evolution	25
	2.4.	Concluding Remarks	27
Chapter 3		Cytogenetics Meets Genomics as a New Area of Biology: Cytogenomics	**29**
Chapter 4		Applied Technologies in the Physical Mapping of Animal Chromosomes	**31**
	4.1.	Routine Methods for Probe Obtaining	32
	4.2.	Probe Labeling and Detection	41
	4.3.	Recent Advances and Application of FISH Technique	43
	4.4.	Microscope Analysis for FISH Assays	47
Chapter 5		Repetitive DNAs: More than just "Junk" DNA	**49**
	5.1.	Basic Features of Repetitive DNAs	49
	5.2.	Multigene Families for Ribosomal RNA and Histones	52
	5.3.	In Tandem Repeated DNAs	55
	5.4.	Transposable Elements	63
Chapter 6		Mapping of Single-copy Genes	**71**
	6.1.	Strategies for Mapping Single-Copy Genes	72
	6.2.	Improving Single-Copy Mapping	75
Chapter 7		Chromosome Painting in Comparative Cytogenetics and Genomics	**79**
	7.1.	Mammals Karyotype Evolution	80
	7.2.	Avian Karyotype Evolution	86

	7.3. The Future of Chromosome Painting	88
Chapter 8	**The "Odd" Chromosomes**	**89**
	8.1. Sex Chromosomes	89
	8.2. B Chromosomes	98
	8.3. B and Sex Chromosomes: Similar or Different?	102
Chapter 9	Final Remarks	**105**
Chapter 10	References	**109**
Index		**153**

Chapter 1

CHROMOSOMES AND GENOMICS

1.1. THE CHROMOSOME HISTORY

The universal occurrence of chromosomes as genetic units of heredity suggests that such cell structures have appeared early in the history of life. The array of genetic material in chromosomes gives several advantages to cells and organisms, packing the genetic information into segregational units and reducing the probability of gain, loss or changes of genetic information. On the other hand, the chromosomes allow for the occurrence of specific and coordinated events of recombination and rearrangement in the genetic material that are of great value to the diversification and evolution of organisms. In this way, the "cytogenetics," which deals with the science of chromosomes, represents a powerful area that aggregates knowledge of cell and molecular biology and can answer several questions on the biology of the species.

The second half of the 19th century is characterized as a remarkable time for genetics and science in general. The mechanisms of heredity began to appear in the works of Gregor Mendel and Charles Darwin, and then a series of significant discoveries involving chromosomes in the end of the 19th century and in the early 20th century allowed for a link between the inheritance of genetic traits and chromosomes. In this way, the cytogenetic science was born with the first analysis of the behavior of chromosomes during cell division, which was carried out at the end of the 19th century by Walter Flemming (Flemming 1882). Flemming provided the first information related to the segregation of chromatin during mitosis in animal cells. At the same time, Eduard Strasburger (Strasburger 1875) observed the mitotic process in plant cells, and further demonstrated that nuclei arise only from pre-existing nuclei. Although Flemming and Strasburger had discovered the chromosomes, the term was coined few years later by Heinrich Waldeyer (Waldeyer 1890). Based on Flemming's discoveries, Theodor Boveri (Boveri 1887) provided the first evidence that the number of chromosomes was reduced in germ cells and restored after the fusion of sperm and egg nuclei. Walther Sutton in 1902 postulated that all chromosomes have a stable structure, or "individuality," that is maintained through generations. With this statement, Sutton articulated the first step of the chromosomal theory of inheritance that was subsequently supported by Thomas Hunt Morgan's experiments with fruit flies, *Drosophila* (Morgan et al. 1915). Morgan made a significant contribution to the science of genetics, stating that genes are located on chromosomes. In the subsequent years the advances on biochemistry would allow for the development of new techniques, contributing to a better characterization of the chromosomal structure. The discovery of the double helix structure of DNA in 1953 (Watson and Crick 1953) gives new directions to the genetics, including cytogenetics. The chromosomes entered a new era with their structure being "molecularly dissected" by advanced molecular biology techniques. The cytogenetics moved from "past cytogenetics," which used a tissue section to identify the human chromosomes in the beginning of 20th century, to the "modern molecular cytogenetics," which utilizes chromosome painting and bioinformatics to recover the evolutionary history of karyotypes. Backing in time we can recover "cytogenetic eras" from "basic cytogenetics" that deals on the description of chromosome number and morphology to the perspective of a "next generation cytogenetics" that will focus in the integration of genomic and nucleotide sequence data to chromosomes (Figure 1.1). The advances in the basic cytogenetic era come from the discovery of the correct diploid number in humans around 1950 (Tjio and Levan 1956) to the development of banding techniques in the 1970's, permitting a more detailed knowledge of chromosomal structure and evolution. After the first use of hybridization of a DNA/RNA molecule to nucleous and chromosomes by Gall and Pardue (Pardue and Gall 1956, Gall and Pardue 1956), several studies were developed, at principle using repetitive DNAs (rRNA genes and satellite DNAs), and latter using fluorescent in situ hybridization (Pinkel et al. 1986) that allowed the simultaneous hybridization of multiple gene/DNA sequences on chromosomes. The studies of Gall and Pardue opened avenues that defined a new era for cytogenetics, the so-called "molecular cytogenetics". The application of bacterial artificial chromosomes (BACs) libraries containing large genomic segments (150-200 kb) and probes containing a DNA pool of whole chromosomes or part of chromosomes, bring up the possibility to refine the chromosome analysis under the focus of "advanced molecular cytogenetis" that allowed the development of chromosome painting and multicolor hybridization in the 1990's. Nowadays, the molecular cytogenetics is in the direction of integrating genomic data to chromosome data using bioinformatic tools for understanding chromosomal structure and evolution. It seems clear that the "next generation cytogenetics" is also going in the direction of integrating other new molecular approaches as microarrays and micro-RNAs analysis to the world of cytogenetics.

1.2. Basic Features of Chromatin and Chromosome Structure

The organization of chromatin and chromosomes represents a general trend in eukaryotes and such a characteristic seems to have arisen early during the evolutionary process of eukaryotic cells. Most of the interphasic chromatin is highly compacted and are structurally inaccessible and functionally inactive. The chromatin has complex hierarchical levels of organization still maintained in the final packing of the chromosomes with several implications to the chromosome manipulation techniques. In this way, the knowledge of chromatin and chromosome organization is critical for the advances in the cytogenetic technologies. Although chromosomes have been well-known for more than 100 years, since the work of Flemming (Flemming 1882), the knowledge of the structural architecture of chromosomes is still under intense discussion. A more organized view of chromatin and chromosomes came from the studies of Rabl (Rabl 1885) and Boveri (Boveri 1888, 1909), who showed that the chromosomes maintain a polarized orientation during interphase, with centromeres and telomeres located on opposite sides of the nucleus, reflecting their conformation during mitosis (the so-called Rabl configuration). The comprehension of chromatin organization developed, mostly in the second half of the twentieth century, with the advances in new technologies of analysis in cell and molecular biology.

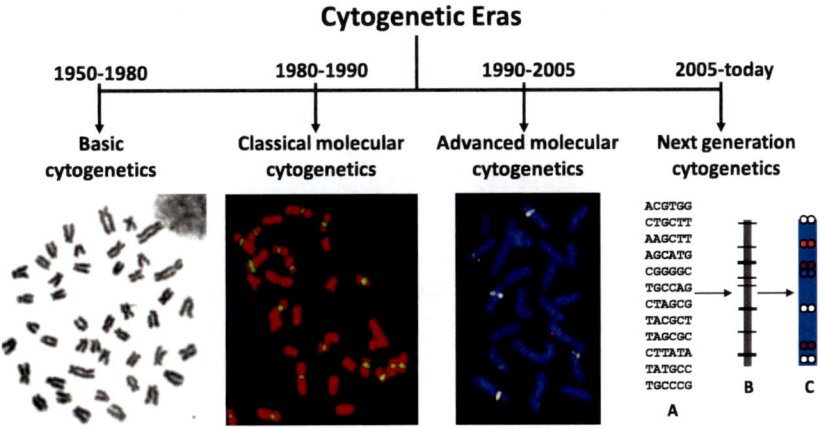

Figure 1.1. Advances in cytogenetics over the years.

1.2.1. Chromatin Structure and Packing

The chromatin can be structurally divided into two types easily visualized in the interphasic nucleus: the "euchromatin" and the "heterochromatin". There are regions of the cell nucleus where chromatin fibers are less densely compacted with a dispersed appearance and less intense color pattern under staining. This material is called euchromatin. On the other hand, some chromatin fibers are densely compacted exhibiting a dense color after staining with appropriated dyes. This material is called heterochromatin. These states represent different degrees of chromatin condensation, reflected primarily by the distinct proteins associated with DNA fibers. In some cases, there is an alternative to the degree of condensation of heterochromatin: some heterochromatin are always compacted and are called "constitutive heterochromatin" whereas another class of heterochromatin, called "facultative heterochromatin", can convert to a kind of euchromatin. The constitutive heterochromatin is often composed of repeated DNA sequences that are primarily not transcribed and the density of genes in this region is reduced, although some transcription degree and presence of genes have been more recently discovered for this genomic regions. The degrees of condensed states of chromatin that define the euchromatic and heterochromatic regions are determined by acetylation and methylation of histone proteins and methylation of DNA. These molecular changes cause the condensation of chromatin fibers and inactivity of genes. Furthermore, specific proteins such HP1 (heterochromatin protein 1) in most eukaryotes, and SIR (silent information regulatory), detected exclusively in budding yeast, are fundamental units of heterochromatin packaging.

There are two types of proteins associated with chromatin: the histones and non-histones. The histones are basic proteins of small size and very conserved among eukaryotes. The non-histones include a large number of diverse proteins of structural and regulatory functions. A more detailed view of chromatin organization comes from electron microscopy analysis that showed chromatin fibers as similar to beads on a string, providing an early clue on the existence of the basic unit of chromatin - the "nucleosome" (for review, Annunziato 2008). The nucleosome core contains eight histone proteins (two units of each H2A, H2B, H3 and H4) that come together to form a discrete protein octamer which binds and wraps about 146 base pairs nucleotide long, that means about 1.7 turns of DNA around the octameric structure of the histones (revised in Wolffe 1999). Another histone, H1 and their variants, binds to the previously described nucleosome core, keeping in place the DNA that was wrapped around the nucleosome. The fact that H1 can be removed easily and does not affect the nucleosome structure suggests its location is external to the particle. Together, the octamer and the H1 histone interact with approximately 160 bp of DNA. The nucleosomes are maintained separated from each other by a segment of variable length from few nucleotide pairs up to about 80. In this way, each nucleosome repeat unit encompasses approximately 200 bp of a DNA segment. Nucleosomes are an invariant component of euchromatin and heterochromatin in the interphasic nucleous, and of mitotic and meiotic chromosomes and represents the first level of chromatin organization. The packing ration of nucleosomes is approximately 6 times fold (Figure 1.2).

The second level of chromatin organization packs the chromatin approximately 40 times fold and is represented by the coiling of the string of nucleosomes into a compacted structure to constitute a fiber of approximately 30 nm in diameter that is observed in both interphasic and mitotic chromosomes (Figure 1.2). The packing of chromatin in the chromosome will be better discussed latter in this chapter. The final level of chromatin organization (third level) involves the packing of the 30-nm fiber giving an overall packing ration of approximately 1,000 times fold in euchromatin and 10,000 times fold in mitotic chromosomes. The heterochromatin has a general packing ration of 10,000, both in interphasic and mitotic chromosomes. Interestingly, histone H1 is very important in stabilizing chromatin higher-order structures, and 30-nanometer fibers are rapidly organized when H1 is present. Changes in chromatin structure are affected by chemical modifications of histone proteins and DNA such as methylation (of DNA and proteins) and acetylation (of proteins), and by non-histone DNA-binding proteins.

The 30-nm chromatin fibers, in specific regions, are attached to the interphasic matrix and chromosome scaffold by proteinaceous structures (Figure 1.2). These regions are called MARs (Matrix Attachment Regions) and SARs (Scaffold Attachment Regions) and seem to be the same DNA region with the ability to compact the chromatin fibers during interphase and cell division, respectively. The MARs DNA regions are usually approximately 70% AT rich, but lacking any consensus sequence. One important characteristic of MARs is the recognition site for topoisomerase II that is also a component of the chromosome scaffold. The presence of topoisomerase II in the MARs suggests some control of topological changes in chromatin operated in this region. In several cases, the same DNA sequences that anchor the nuclear matrices are also observed in the metaphasic scaffold. Different proteins can be found in the nuclear matrix and in the chromosome scaffold, although there are some common components, such as topoisomerase II. In fact, MARs and SARs proteins are characterized by their preparation from interphase or metaphase cells, respectively. During interphase, the nuclear matrix represents the counterpart of the chromosome scaffold of mitotic chromosomes.

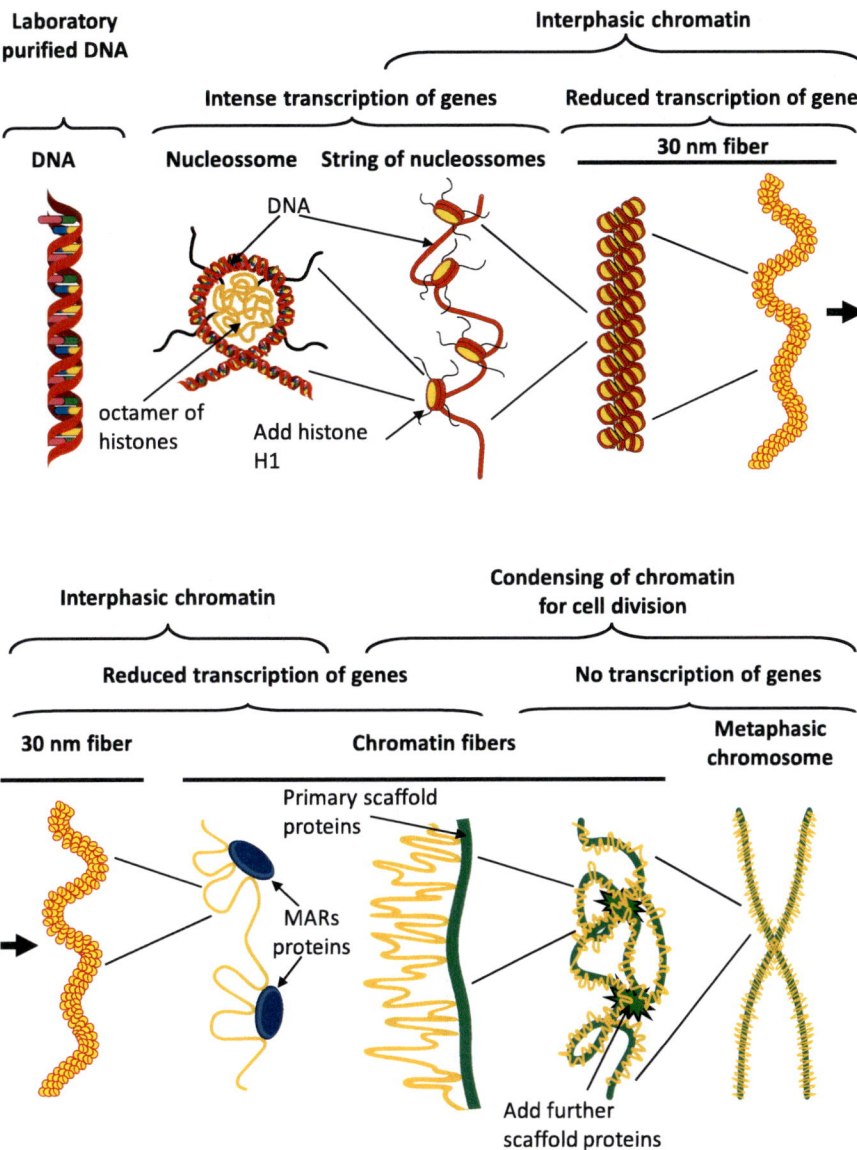

Figure 1.2. Levels of chromatin organization in eukaryotic cells.

1.2.2. The Eukaryotic Chromosome: Highly Compacted Chromatin

The eukaryotic chromosomes are composed of DNA associated to proteins as described in the beginning of this chapter and represent the super coiled state of the nuclear chromatin that appears during mitosis and meiosis. Several models have been proposed for all states of higher condensation of chromatin in the chromosomes (Revised in Wanner and Formanek

2000): (i) folding of the 30-nm fibers in a looping manner to form chromosomes without intermediate levels of packing; (ii) supercoiling of the 30-nm fibers by a hierarchy of helices; (iii) anchoring of loops formed by the 30-nm fibers to an axial fibrous element called the scaffold; (iv) **packing of DNA in "chromomeres" (coiled solenoides)** and a dynamic matrix formed of parallel protein fibers. However, there are several remaining questions concerning chromatin organization in chromosomes and this topic is still under discussion.

During the condensation of interphase chromatin into mitotic and meiotic chromosomes, the nuclear matrix is converted into more substantial chromatid cores by adding proteins and/or modifying proteins of the matrix. There have been a number of different proposals (sometimes conflicting) to explain coiling during chromosome condensation. The highly organized ultrastructure of mitotic chromosomes is composed of approximately 60 kb loops of DNA that are anchored in a central proteinaceous scaffold. The condensation of chromosomes begins in early cell division and is coordinated by the machinery involved in the cell cycle control. At this stage, the gene expression shuts down and the histones suffer modifications such as acetylation and phosphorylation of specific amino acid residues leading to the reorganization and compaction of chromatin. The chromosome compaction is also helped by "condensins", a class of large protein complexes belonging to the SMC (structural maintenance of chromosomes) proteins. The cell cycle control complex ciclin-Cdk1 acts in the phosphorylation and activation of condensins. Although itis not known how condensins act on the condensation of chromosomes, when added to isolated DNA in the presence of ATP, condensins organize large loops of DNA.

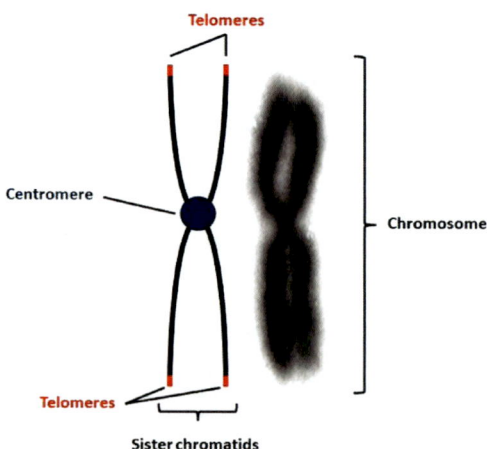

Figure 1.3. Mitotic metaphasic chromosome structure.

The essential feature of all chromosomes is represented by centromere and telomere (Figure 1.3). The centromeric region, also called as primary constriction is a region where the sister chromatids are held together along the cell division until the anaphase. This region contains the kinetochore, which is responsible to attach the chromosome to the microtubules of the mitotic spindle, and regulates the proper pairing and segregation of the chromosomes during cell division. Despite their conserved, essential function, centromeres are characterized by a dynamic evolutionary region, both in relation to DNA and protein contents (Henikoff et al. 2001). Although they have been well investigated from animals to fungus, important points remain to be understood. Centromeric function in *Drosophila* has been assigned to a 200-600 kb region and seems to contain several dispersed elements, including sequences required for kinetochore assembly and sister chromatid pairing, among others. Disturbances in the structural and functional organization of the centromeres are critical leading to problems such as development defects and cancer. The centromeric regions are rich in repetitive DNAs, which is a trait of the characterized centromere regions from humans, mice, maize, fruit fly, and yeast (Henikoff et al. 2001). An extensive analysis of centromeric satellite DNAs of vertebrates showed the presence of short A-rich motifs, typical characteristics of centromere satellite (Vinãs et al. 2004).

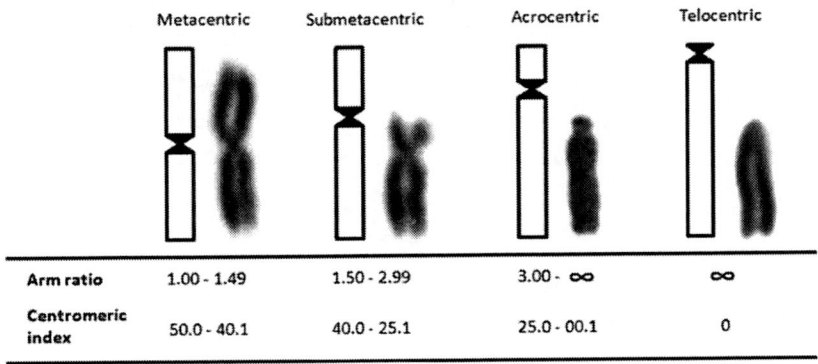

Figure 1.4. Centromere position and chromosomal shape with limit values for each chromosome type.

Another important region of all chromosomes is the telomeres that signalize the "end" of the chromosome. In a general way, the telomeric region of chromosomes contains structural features that confer (1) stability and (2) maintenance to the linear DNA molecule. Telomeric DNA sequences have been characterized from a wide range of eukaryotes. Similar sequences are found in distant organisms as plants and humans suggesting the telomere organization and function follow a universal principle. Each telomere consists of series of 100-1000 short repeats tandemly organized, which can be written in the degenerated form as $C_n(A/T)_m$, where $n>1$ and m is 1-4. There are two important organization features of the telomeric DNA: (i) the telomeric DNA seems to be organized *in vitro* in an atypical pattern of association between the G residues of the short-tandem repeats forming a helical structure; (ii) the 3' single stranded end of the telomeric DNA displaces the same sequence in an upstream region of the telomere forming a loop. Both structures "seal" and protect the end, giving (1) stabilization to the chromosome. Besides the stabilization and protection of the chromosomes, the telomere also needs to be (2) extended to solve the problem caused by the shortening of DNA at the chromosomal end after each cell replication cycle. This is caused because the DNA replication machinery does not duplicate the very end of the chromosome. There are several proteins involved in the extention of telomeric DNA, one being the "telomerase", which needs to be mentioned. The telomerase is an example of a reverse transcriptase, an enzyme that synthesizes DNA sequence using an RNA molecule as a template. There is a short RNA component in the telomerase that includes a 15-22 nucleotide long that is identical to the tandem repeats of the telomeric DNA. This short RNA segment acts as a template for the synthesis of telomeric repeats, without the template of the cell DNA.

1.3. A Brief History of Karyotypes under the Focus of Basic Cytogenetics

Since the first observation of chromosomes under the microscope, the methodologies used for chromosomal obtaining and karyotype analysis have been evolved, permitting a better knowledge about chromosomal structure, DNA content, behavior and evolution. Although these advances have occurred, the basic features of karyotypes has been observed under the conventional staining, and for a long time this approach has permitted the description of diploid number, chromosomal morphology and size (Figure 1.4), fundamental number of arms (FN) and the presence of sex chromosome mechanisms in diverse animal groups. Furthermore, the conventional staining even allowed the description of chromosomal polymorphisms, including heteromorphic chromosomes, karyomorphs, and the presence of supernumerary elements.

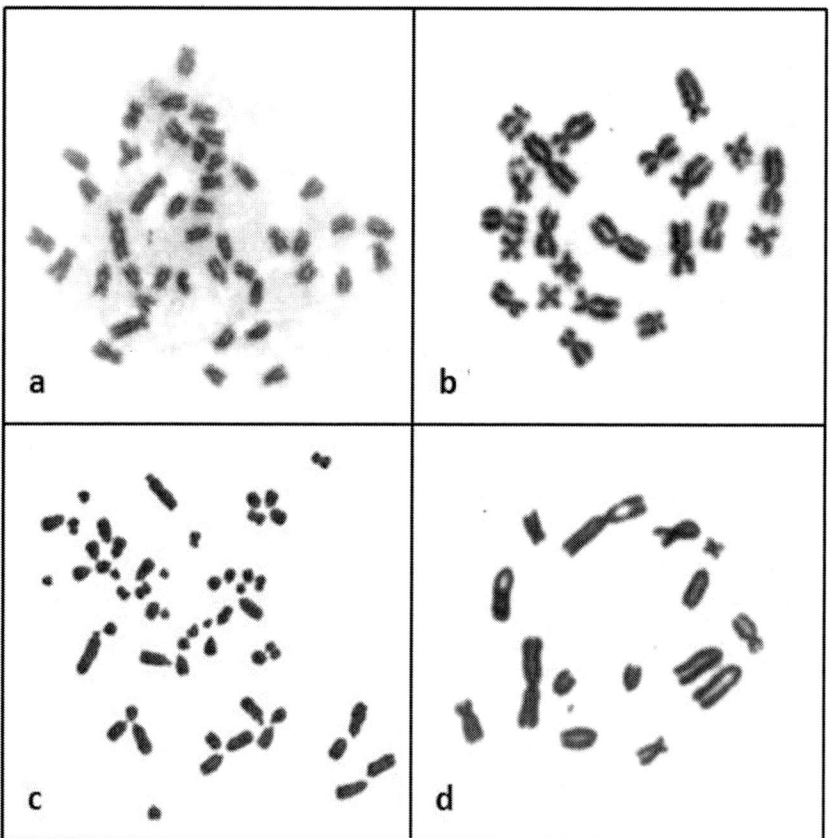

Figure 1.5. Mitotic spreads of vertebrate chromosomes. (a) *Metriaclima lombardoi* (Cichlidae, fish) 2n = 44; (b) *Physalaemus cuvieri*, (Leptodactylidae, amphibian) 2n = 22; (c) *Myiozetetes cayanensis*, (Tyrannidae, bird) 2n = 56; (d) *Tonatia bidens*, (Phyllostomidae, mammal) 2n = 16,XY. Note the presence of large chromosomes in (b) and microchromosomes in (c), noteworthy characteristics for amphibian and bird karyotypes, respectively. The figures "b", "c" and "d" are courtesy of Lourenço LB (Universidade Estadual de Campinas, Brazil), Oliveira EHC (Universidade Federal do Pará, Brazil), and Santos N and Souza MJ (Universidade Federal de Pernambuco, Brazil), respectively.

The conventional analysis involving the description of karyotypes and chromosomal behavior are performed mainly by the analysis of mitotic and meiotic chromosomes, and for specific groups, such as Diptera, also by the observation of polytene chromosomes. In general, for some groups the studies have been conducted through the observation of mitotic chromosomes, such as vertebrates (Figure 1.5). On the other hand, for invertebrates, such as beetles, grasshoppers, bugs, mollusks and others, obtaining the mitotic chromosome is more difficult and the studies can also be conducted by analysis of meiotic elements (Figure 1.6). Independent of the chromosome obtaining method, this conventional analysis had revealed diverse chromosomal patterns (Figures 1.5 and 1.6). In fact, the general structure of chromosomes is conserved among animals, and most representatives of this group presents monocentric chromosomes that can vary in number, size and morphology. However, some specific groups present chromosomes that show kinetic centromeric activity along a part or in its entire length, being called holocentrics or holokinetics, and are observed, for example, in some arthropod groups, i.e. insects and arachnids.

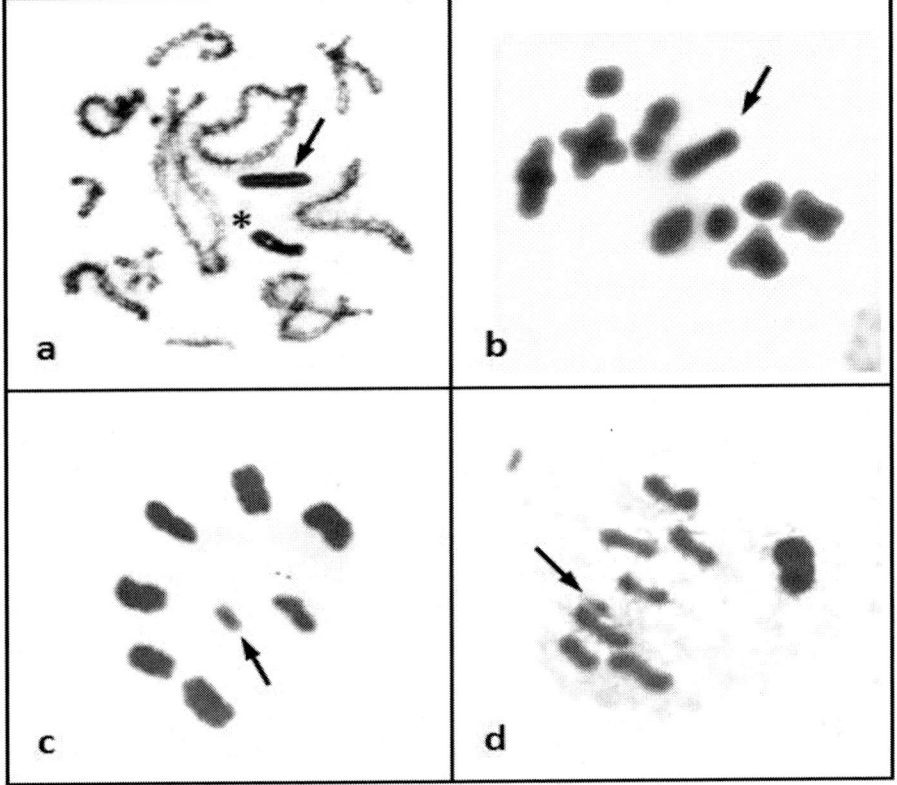

Figure 1.6. Meiotic cells of insects. (a) Diplotene from *Rhammatocerus brasiliensis* (Acrididae, Orthoptera) 2n = 23,X0 + B, (b) Metaphase I from *Stiphra robusta* (Proscopiidae, Orthoptera) 2n = 19,X0; (c) Holocentric chromosomes in Metaphase I from *Edessa rufomarginata* (Pentatomidae, Heteroptera) 2n = 16,XY; (d) Metaphase I from *Dichotomius bos* (Scarabaeidae, Coleoptera) 2n = 18,Xy_p. The arrows indicate the sex chromosomes and the asterisk in (a) one B chromosome. The figures "a" and "c" are courtesy of Loreto V and Souza MJ (Universidade Federal de Pernambuco, Brazil), respectively.

Although the basic chromosome structure is conserved, they are not "static" and there are several types of changes along the species diversification history regarding number, morphology, size and DNA content of the chromosomal elements. The mechanisms of changes in chromosomes are defined as "chromosomal rearrangements", being the most common macro-chromosomal changes in the animal kingdom the translocations, inversions, fissions and fusions, besides the duplications, deletions and transpositions (Figure 1.7), which can occur between homologous or non-homologous elements (White 1973, King 1993). These rearrangements have caused the existence of highly differentiated sets of chromosomes in all aspects, as number, size and morphology, which, in general, are more similar in related species than in distantly related species, although there are some examples of related species with extremely distinct karyotypic structures. For vertebrates and in the lesser extent for invertebrates, mainly in insects, there are good examples of conserved and variable karyotypes between related species and also the presence of particular karyotypic structure and chromosomal changes commonly observed in some groups. Some of these examples will be briefly presented here and more detailed information can be found in specialized journals, although some of this data is out of age and fragmented due the new technologies applied in chromosomal studies (see the chapters IV, VII and VIII).

1.3.1. Chromosomal Variability Revealed by Standard Staining

Perhaps the most famous example of chromosomal differentiation between related species is that of the species *Muntiacus reevesi* (Chinese muntjac) and *M. muntjac vaginalis* (Indian muntjac), in which the Chinese muntjac has 2n = 46 chromosomes (a typical diploid number in mammals) with acrocentric morphology and the Indian muntjac has only 2n = 6 or 7 chromosomes in female and male, respectively (Wurster and Benirschke 1967, 1970). Using the classical analysis, it was defined that the Indian muntjac karyotype probably results from multiple tandem fusions and few centromeric fusions of ancestral acrocentric chromosomes with smaller size (Hsu et al. 1975). This hypothesis was confirmed by some works using differential techniques and molecular cytogenetic approaches that indicated the occurrence of other chromosomal rearrangements besides the fusions (Shi et al. 1980, Elder and Hsu 1988, Yang et al. 1997a). In other groups, variations have been frequently described, in both more inclusive (such as orders, families) and less inclusive (such as genus) taxa, besides the variation in the species level, characterizing chromosomal polymorphism among populations.

For Neotropical fishes the diploid number ranges from 2n = 20 to 134 with the presence of some cytogenetically variable groups which present high variability concerning diploid number, karyotipic formulae, presence of sex chromosomes and supernumerary elements, although it is possible the identification of specific patterns for distinct groups (Almeida-Toledo et al. 2000, Oliveira et al. 2007). Examples of this intense variation can be observed in the order Characiformes that possesses diploid numbers ranging from 2n = 28 in *Hemigrammus* to 2n = 102 in *Potamorhina altamazonica*, in contrast with the modal number 2n = 54 chromosomes of the group (Reviewed by Almeida-Toledo et al. 2000). In Gymnotiformes, a wide variation was detected in the less inclusive taxon, as observed in the representatives of *Gymnotus* with 2n = 39/40, 40, 42, 48, 50, 52 and 54. The sex systems are also extremely variable in fish and although some species do not present sex differentiated chromosomes as observed in *Gymnotus*, eight types of sex determining systems have already been observed, involving non- differentiated chromosomes, presence of XX:XY or XX:ZW systems with simple or multiple chromosomes (Almeida-Toledo et al. 2000, Oliveira et al. 2007). An interesting example of presence and variability of sex chromosomes was related in the species *Hoplias malabaricus* (Characiformes) with the occurrence of XY, X_1X_2Y and XY_1Y_2 sex chromosome system found in karyomorphs B, D and G, respectively. Other variations at intraspecific levels were also described for this species, and until now, seven karyomorphs (A-G) were identified, presenting variations in the morphology, size, diploid number and presence/absence of sex chromosome systems (Bertollo et al. 2000).

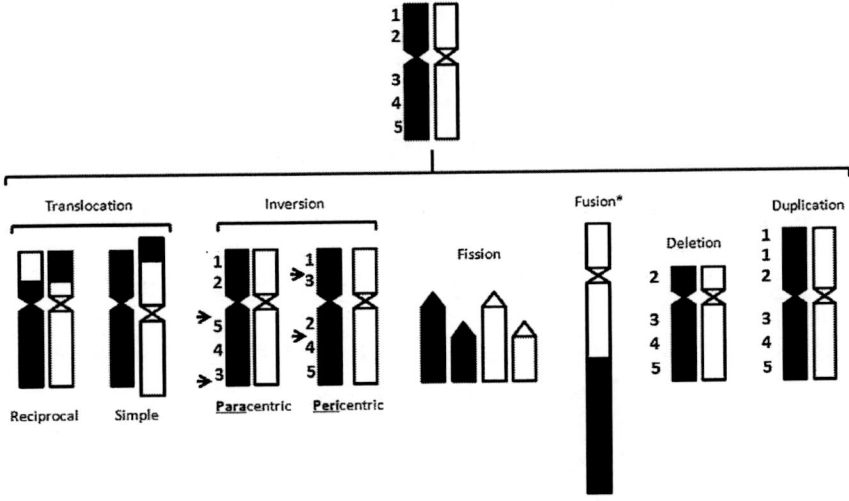

Figure 1.7. Representation of some common chromosomal rearrangements in animal chromosomes. * Telomere-telomere fusion (*In tandem* fusion). The fusions can also involve the centromeric region (Robertsonian fusion).

In the opposite way of the chromosomal variability trends there are examples of fish groups presenting an evolutionary karyotypic stasis or few occurrences of large chromosomal changes, as observed, for example, in some marine fish groups, such as Scianidae and Serranidae families (Galetti Jr et al. 2000, Molina 2006). For freshwater fish, the occurrence of conserved karyotypes, at least related to the diploid numbers, can be observed in representatives belonging to the family Cichlidae (Feldberg et al. 2003, Poletto et al. 2010a). Among Scianidae species, the presence of 2n = 48 chromosomes with only one arm, FN = 48, is a remarkable characteristic, being this condition is regarded as a primitive cytogenetic feature widespread in karyotypes of Perciformes, including several families and suborders (Galetti Jr et al. 2000, 2006, Molina 2006). According to Molina (2006), the reduced rate of chromosomal divergence could be explained by the high dispersal capacity of larvae and connectivity among distant populations, due to the presence of less effective biogeographic barriers in the marine environment. This chromosomal structure has resisted along speciation if compared with morphological divergence. Concerning cichlid fishes, the presence of 2n = 48 modal number in the American species possibly reflects the same condition of Perciformes, and for the African ones, the diploid number 2n = 44 is a common characteristic. In spite of this conservation of the diploid number, small differences related to the number and chromosomal morphology are frequent as a consequence of specific chromosomal rearrangements (Feldberg et al. 2003, Poletto et al. 2010a).

For the Amphibian class there are chromosomal studies, for example, in representatives of Caudata (salamanders) and Anura orders. The majority of salamander families (around 60%) presents conserved karyotypes regarding diploid numbers, with 2n = 22-28, and the presence of bi-armed chromosomes (Morescalchi 1973, 1975, Green and Sessions 1991, 2007, Sessions 2008). For representatives belonging to the families Hynobiidae, Cryptobranchidae, Sirenidae and Proteidae (Caudata), higher diploid numbers have been described, with 2n = 40-78, 2n = 60, 2n = 46-64 and 2n = 38, respectively, besides diverse chromosomal size and morphology (Reviewed by Sessions 2008). The model of chromosomal evolution for salamanders proposed by Morescalchi (1973, 1975) considers the telocentric morphology as a plesiomorphic condition, while the bi-armed chromosomes take place as a derived state. Although a uniform karyotypic structure was seen in salamanders, the presence of chromosome variation have been reported, such as somatic polymorphism, presence of B chromosomes and polyploidy (Green and Sessions 2007). According to Sessions (2008), it is possible that polyploidy played a major role in the chromosomal evolution in this group. Concerning sex chromosomes, differentiated sex chromosomes were reported in approximately 50 representatives belonging at least to 15 genera and six salamanders families, including XX:XY and ZZ:ZW systems, being the XX:XY the most common (~90%) (Green and Sessions 2007, Sessions 2008). According to Sessions (2008), other examples of heteromorphic sex chromosomes will be discovered among salamander species with the use of modern cytogenetic techniques.

In the Anura order, that presents the highest diversity for the Amphibian class with about 90% of the species, the chromosomal characteristics have been used for studies of phylogenetic relationship, systematic and karyotypic differentiation among related and non-related groups. Most of anurans possess a low chromosomal diploid number with large meta-submetacentric chromosomes, being the diploid number 2n = 26 considered ancestral for some groups (Bogart 1973, King 1990). In this sense, for a long time it has been assumed that the anurans have a conservative karyotype structure. On the other hand, this conservatism is not observed for all groups and, for example, in the frogs some variation have been reported (Kuramoto 1990, Bogart 1991). According to Bogart (1991), the chromosomal differentiation in anurans apparently is slower compared to other vertebrate groups. Morescalchi (1968, 1973) and Lynch (1971) suggested that anurans with primitive morphological characteristics present telocentric chromosomes and some authors have suggested that in tandem fusions were the principal mechanism acting in the reduction of the diploid number in this group (Bogart 1973, 1991, Beçak et al. 1970, Morescalchi 1973, Miura 1995). Furthermore, there is not enough information to point the direction of chromosomal transformation being impossible to determinate if the changes occurred by decrease or increase in the diploid number (Bogart 1991).

For reptilians, the karyotypic constitution has been studied in species belonging to four orders of the group: Testudines (turtles), Crocodilia (crocodiles and alligators), Squamata (lizards and snakes) and Rhynchocephalia (tuataras). The diploid number in turtles ranges from 2n = 26 to 96, being that this variability results from the differences in the number of microchromosomes (Bickham 1984, Bickham et al. 1985, Olmo 1986). Crocodilia presents diploid number varying from 2n = 30 to 42, meta- and submetacentric chromosomes and telo-subtelocentric chromosomes (Cohen and Gans 1970). The fundamental number varies from

56 to 62 mainly without sex chromosomes heteromorphism and absence of microchromosomes. For the most recent reptilian clade, the Squamata order, which also represents the most specious group with over 6,000 species described, the chromosome complement is conserved for some groups in contrast with variability related for the others. In snakes, the chromosome complement is relatively conserved, being the most common chromosomal number 2n = 36, with the presence of 16 macro- and 20 microchromosomes varying mainly in size (Beçak et al. 1964, Beçak and Beçak 1969, Singh 1972, Olmo and Signorino 2005, Matsubara et al. 2006, Oguiura et al. 2010). In contrast with this condition, the lizards (Lacertilia) have showed large variation in the chromosome number and morphology, and they generally exhibit two distinct karyotypic conditions. One is composed by macro- and microchromosomes, with clear size distinction, and another one is composed by chromosomes with similar sizes, presenting few or no microchromosomes (Gorman 1973, Olmo 1986, Schimid et al. 1994, in den Bosch et al. 2003, Olmo and Signorino 2005).

Around 800 karyotypes have been published for birds and the main characteristic of chromosomal structure for this group is the presence of macrochromosomes, several microchromosomes (characteristic also observed in reptilians) and ZZ:ZW sex system (Rodionov 1997). The diploid number varies from 2n = 40 in *Burhinus oedicemus* and *Ceratogyman bucinator* to the highest 2n = 136-142 in *Corythaixoides concolor* (Christidis 1990). The diploid numbers range from 2n = 74 to 86, occurring in most part of this group (around 63%), and from 2n = 66 to 74, observed in approximately 24% of the species studied (Christidis 1990). In comparison with the variability of mammal karyotypes, the avian ones show lower variation and according to Griffin et al. (2007), this pattern could represent a successful karyotype organization regarding the genome structure in the group.

Undoubtedly, the most part of consistent and elucidative studies about animal karyotype evolution were performed in mammals, and more than 1,000 species of this group were cytogenetically described. By comparison with other vertebrate clades, mammals display a high degree of karyotype variability, presenting groups with extensive variation in the karyotypes, although more conservative groups have also been described. Since the correct description of human diploid number by Tjio and Levan (1956), chromosome numbers were studied in some mammal groups, such as primates, rodents, marsupials and others (Hsu and Benirschke 1967, O'Brien et al. 2006) and the chromosome number and morphology were noted to be a distinct feature for most species. On the other hand, some groups such as Felidae present similar karyotypes (Wurster and Benirschke 1968). Chromosome numbers for mammals vary from 2n = 6 in the Indian muntjac to 2n = 102 in the red viscacha rat (Wurster and Benirschke 1970, Contreras et al. 1990) with distinct chromosomal morphology, which can be mainly acrocentric or a combination of metacentric, submetacentric and acrocentric chromosomes, generating distinct fundamental numbers. These standard analyses in the mammalian chromosomes, as in other animal groups, pointed primarily to the 2n and chromosomal shape and using these information different hypotheses were proposed about chromosomal evolution in this group. Ohno (1970) proposed the "fission hypothesis" in which the ancestral karyotype for mammals consisted by 2n = 96 and acrocentric chromosomes, and the karyotypes of the present day species arose from a variable number of metacentric fusions. On the contrary, Todd (1970), influenced by studies in marsupial chromosomes, suggested the ancestral chromosomal structure for mammals composed by 2n = 14 large chromosomes postulating the "fission hypothesis". Finally, according to Matthey (1973), the ancestral karyotype consisted most probably by 2n = 48 meta-, submeta- and acrocentric elements. In fact, the chromosomal studies advanced more rapidly for mammals and the chromosomal comparison using only standard staining takes an out of age place for some groups. Modern cytogenetic techniques, such as whole chromosome paint, gave new contributions about the evolutionary relationship between a great number of mammalian species that will be discussed in Chapters IV and VII.

For invertebrates, the chromosomal studies have mostly been developed in insects, with most of the analysis performed in specific groups as Coleoptera, Hymenoptera, Hemiptera, Homoptera, Heteroptera, Diptera, Orthoptera and Lepdoptera. Although insects correspond to the most diverse animal group and have high diversity of karyotypes, including the presence of diverse diploid numbers in related species, absence or the presence of different sex chromosome systems, presence of parasitic elements, occurrence of monocentric and holocentric chromosomes, presence of chiasmatic and achiasmatic chromosomal behavior, and inverted meiosis in some groups, the cytogenetic studies in this class are still scarce compared to this cited diversity.

In fact, it is almost impossible to define the chromosomal evolutionary pathways for insects as a whole, due to the extensive variability in their chromosomes. Although, remarkable characteristics for specific groups can be described using standard analysis. For example, in grasshoppers belonging to Acrididae family, a high conservation in the karyotype structure with $2n = 23$, acrocentric chromosomes and X0:XX sex system is observed, and only in specific groups extensive variation was reported. For some genera of Gomphocerinae grasshoppers, $2n = 17$ chromosomes was described, being that this structure originated from three centric fusions, corroborated by the presence of three large biarmed chromosomes (meta- or submetacentric) (Hewitt 1979). In the Neotropical Melanoplines grasshoppers, mainly due the occurrence of Robertsonian translocation, the presence of neo-XY sex system is common and more than 75% of the species belonging to the tribe Dichroplini have derived karyotypes (Mesa et al. 1982).

For beetles, besides the variability in the chromosomal number, an intriguing variation occurs in relation to sex chromosomes. According to Smith (1953), the sex mechanism in beetles can be divided in six groups related to their structure and mode of association in meiosis, X0, Xy_p, Xy_r, Xy_c, neo-XY and "Xy", besides other variations in each mechanism and the presence of multivalent, with chiasmatic or achiasmatic behavior. Many characteristics make heteropterans unique among insects. The representatives of this group present holokinetic chromosomes (without localized centromere); presence of distinct meiotic behavior for autosomes and sex chromosomes, being characterized by a pre-reductional type of meiosis in which the autosomal bivalents segregate reductionally, although the sex chromosomes are achiasmatic and divide equationally at the first meiotic division. The second meiosis is equational for autosomes and reductional for sex chromosomes (Ueshima 1979). Moreover, for some families the presence of m-chromosomes is remarkable. These chromosomes are small in size, usually unpaired during early meiosis, coming closely associated together in late diakinesis, showing reductional segregation at anaphase I (reviewed by Papeschi and Bressa 2006). The holokinetic chromosomes observed for all heteropteran species are also noteworthy in some representatives from Lepdoptera, Homoptera and Hemiptera (reviewed by Smith 1960).

The characteristic of most hymenopterans (bees, ants and wasps) is the presence of a diploid chromosomal set in females and haploid in males, called haplodiploid sex determination system. In this system, sex is determined by the number of chromosomal sets received by the individual, in which a fertilized egg by a sperm generated females (diploid), while unfertilized eggs generate males (haploid) (reviewed by Heimpel and Boer 2008). In dipterans, the remarkable characteristic is the presence of polytene chromosomes. These chromosomes facilitate the cytogenetic analysis and interpretation of chromosomal rearrangements, due the characteristic light and dark bands along their extension. They were observed for the first time in the salivary gland of *Chironomus* midges by Balbiani (1881), and detailed maps of dipterans species are still widely used today in classical cytogenetics, primarily for characterization of interpopulational polymorphism and phylogenetic studies, besides their use for genome organization and for anchoring molecular maps.

Besides this variability in the normal complements (A complement), the standard analysis has also revealed the presence of supernumerary chromosomes in some animal groups. In eukaryotes as a whole, the presence of B chromosomes were detected in about 15% of the species, including around 500 animal species (reviewed by Camacho 2005). This standard analysis has revealed distinct types of B elements with variable morphology, size and number. For example, fish B elements have been described frequently in distinct species and populations with variable number and sizes. The B elements can vary from punctiform to large B elements, for example in *Alburnus alburnus*, that harbor the largest B chromosome for vertebrates, corresponding to around 10% of its genome (Camacho 2005, Schmid et al. 2006). For amphibians, about 27 species harbor B elements, including one anuran from New Zealand that presents one of the largest variation of B chromosome numbers for vertebrates, in which individuals with up to 16 B elements have been observed (Green 1988). The higher number of B elements was described in mice belonging to the family Muridae, *Apodemus peninsulae*, with up to 24 B elements with high variability related to frequency, size, morphology in intrapopulational and interpopulational level, besides the presence of intra-individual mosaicism (Kartavtseva and Roslik 2004). On the contrary of this widely distribution of B elements in vertebrates, for reptiles and birds this polymorphism was observed in few species, being about 15 in reptilians and only one in birds (Reviewed by Camacho 2005). This low B chromosome detection in these groups could be related to the presence of microchromosomes, that makes difficult the analysis conducted by standard staining.

For invertebrates, the B elements were described, for example, in Mollusca, Platyhelmintes and Arthropoda (mainly insects). In insects, the most part of species with B elements belongs to the order Orthoptera, followed by Coleoptera and Diptera, with about 210, 69 and 62 species harboring this polymorphism, respectively (reviewed by Camacho 2005). The B chromosomes were first discovered in insects more than 100 years ago in the leaf-footed plant bug from the genus *Metapodius* (Wilson 1907a, b). The variation in number of Bs in distinct species is from one to 20, being the s largest number observed in the dipteran *Xylota nemorum* with $2n = 10 + 20$ B elements. Examples of well-known B chromosome systems were reported in the grasshoppers, as for example in *Eyprepocnemis plorans*, and in fact, the B chromosomes in grasshoppers are relatively well-studied in comparison to other invertebrate groups (reviewed by Camacho 2005). According to Camacho (2005), this prevalence and well-studied B chromosome for specific groups is probably a by-product of higher intensity of studies, which sometimes can reflect in part the facility of chromosome obtaining and analysis.

1.3.2. The Chromosomal Banding Advent

Although for long time the chromosomal aspects and karyotype evolution were studied under the focus of standard analysis, in the 1970s, the chromosomal banding techniques became available (Figure 1.8, 1.9). The several banding techniques can be produced by different dyes or treatments and reveal finer aspects about the chromosome structure and organization permitting a more detailed comparison between animal karyotypes and the ability to hypothesize more precisely the presence of homologous segments. Among these differential banding techniques, the most common used for animal chromosomes are the C-banding, silver nitrate staining (using $AgNO_3$) and G-banding (Figures 1.8a-h, 1.9b-d). The first two techniques are useful for comparative analysis concerning the repeated DNAs, with the C-banding used mostly for heterochromatin analysis (Sumner 1972), while the silver nitrate staining is useful for determination of active ribosomal sites, named Nucleolar Organizer Regions (NORs) (Sumner 1990).

The C-banding is produced by a treatment with diluted acid, alkali, warm saline solution and stained with Giemsa dye and reveals most constitutive heterochromatin (Figure 1.9b) (Sumner 1972). These heterochromatic blocks can occur in every part of chromosomes, although in general, they are located preferentially in specific chromosomal parts, such as in the centromeric region and in the lesser extend in the terminal regions, or in specific chromosomes, for example, autosomes, sex elements or supernumerary elements. These heteorochromatic segments may vary in quantity and size in the karyotypes, from very small to large blocks, although theygenerally present equilocal positions. In fact, this variability related to the quantity and size of the C-banded blocks has been observed in intra-individual, intra-specific and inter-specific levels generating heteromorphism and polymorphism in a range of animal groups. Among the applicability of the C-banding for animals, there are some examples of the use of this technique as an important tool for chromosome identification, identification of markers to distinguish homologues and in studies of chromosomal evolution, being for some groups the main strategy for these purposes. In general, this technique has been useful for the distinction of specialized chromosomes, such as sex chromosomes and supernumerary elements. Although the C-banding has been used as a universal tool in animal cytogenetic, the most interesting examples of applicability of C-banding for chromosomal distinction and karyotype diversification have been published in insects, fish, amphibians and mammals.

The silver nitrate staining ($AgNO_3$) banding technique detects the active NORs in distinct cell cycle phases, being the principal method for identifying sites of ribosomal DNA (rDNA) on chromosomes since 1980 (Figure 1.9c) (Sumner 1990). Although this technique remains in use nowadays, it has been replaced by the mapping of rRNA genes using the *in situ* hybridization, that is more specific and confinable (Sumner 1990, 2003) (see chapter IV). Frequently, in species with multiple NORs, not all are stained, due the restricted activity of the sites of these elements. In fact, the silver nitrate stained sites correspond to the rDNA sites that presented activity during the preceded interphase, due the affinity of $AgNO_3$ to the acid proteins that are associated to rRNA produced by the rRNA genes (Sumner 1990; Wachtler and Stahl 1993). This technique has been used and has permitted the accumulation of a huge amount of information about this chromosomal marker (NOR), that is useful for the characterization of general structure of karyotypes and contributed to the knowledge of

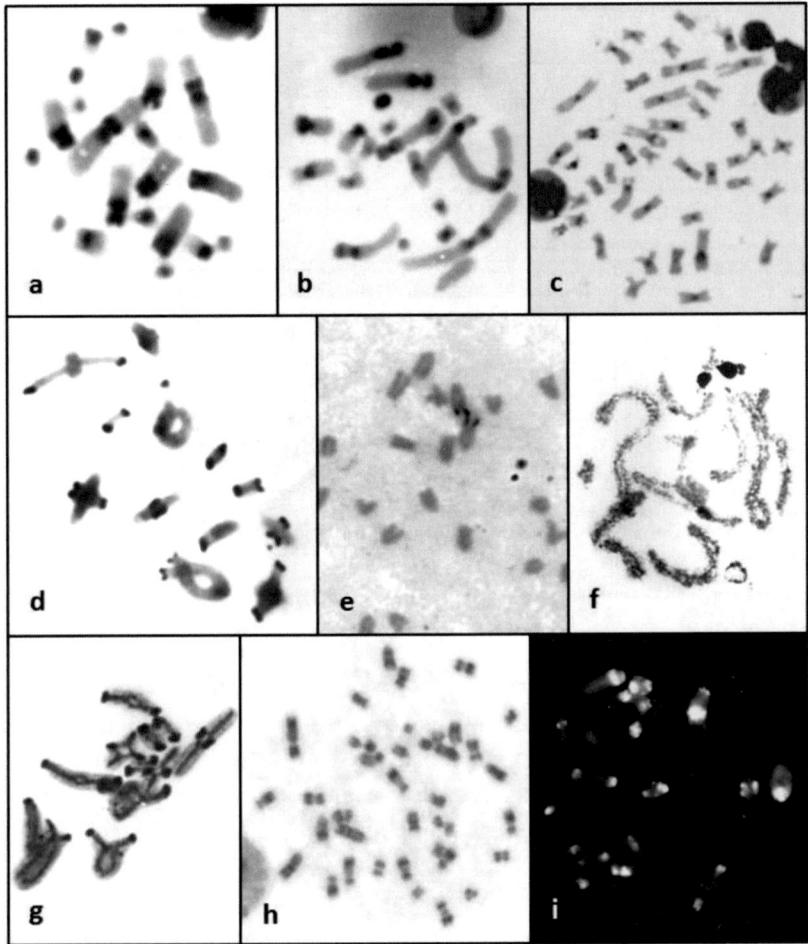

Figure 1.8. Differential chromosomal staining of mitotic vertebrates chromosomes (a-c,e,h) and meiotic invertebrates (f,g,i) chromosomes. (a,b) C-banded chromosomes from *Carollia perspicillata* (Phyllostomidae, mammal) (a) female 2n = 20,XX and (b) male 2n = 21,XY_1Y_2; (c) C-banded chromosomes from *Hoplias malabariicus* (Erythrinidae, fish) 2n = 42; (d) C-banding in Metaphase I from *Xyleus descoideus angulatus* (Acrididae, Orthoptera) 2n = 23,X0; (e) silver nitrate staining in chromosomes from *Cichla kelberi* (Cichlidae, fish) 2n = 48; (f) silver nitrate staining in Pachytene from *Agriacris auripennis* (Romaleidae, Orthoptera) 2n = 23,X0; (g) silver nitrate staining in Metaphase I from *X. d. angulatus* (Acrididae, Orthoptera) 2n = 23,X0 showing the affinity of silver nitrate for the chromosomal core and kinetochores; (h) G-banded chromosomes of *Sus scrofa domesticus* (Artiodactyla, Mammalia) 2n = 38,XY; (i) CMA_3/DA fluorochrome staining in Metaphase I of *X. d. angulatus* (Acrididae, Orthoptera) 2n = 23,X0 showing GC rich regions. The figures "a" and "b" are courtesy of Santos N and Souza MJ (Universidade Federal de Pernambuco, Brazil); figures "d", "f", "g" and "i" are courtesy of Souza MJ (Universidade Federal de Pernambuco, Brazil); figure "c" is courtesy of Venere PC (Universidade Federal de Mato Grosso, Brazil); figure "h" is courtesy of Mota L (Universidade Estadual Paulista, Brazil).

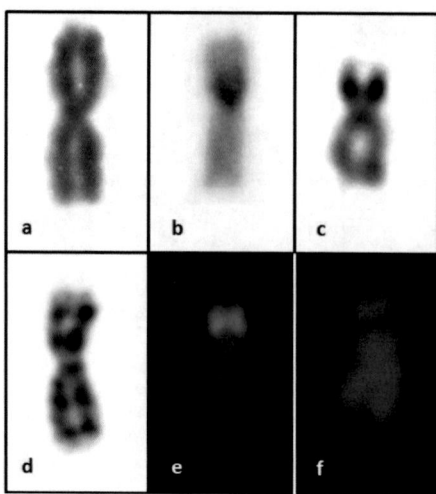

Figure 1.9. Detail of standard staining and distinct chromosomal banding of vertebrate chromosomes. (a) standard staining; (b) C-banding; (c) silver nitrate staining; (d) G-banding; (e,f) sequential CMA_3 and DAPI fluorochrome staining, respectively. Note in (e,f) that the positive CMA_3 bands correspond to negative DAPI staining.

chromosome evolution. The NORs have been used for the characterization of intra-specific polymorphism among distinct populations of some species. In fish and amphibians, the NORs investigation allowed comparative analysis between related species indicating specific chromosomal rearrangements, and were also useful for the determination of activity of B chromosomes, as described in grasshoppers. Eventually, this technique is available for staining other chromosomal regions, such as telomeres, kinetochores, cores and heterochromatic regions (Figure 1.8g). The affinity for these other chromosomal regions sometimes makes the analysis difficult, making this banding approach useless for some groups.

The G-banding produces a characteristic pattern of alternating light and dark regions (bands), generating a longitudinal differentiation of euchromatin along the chromosomes (Figure 1.9d). It has been applied for recognition of homologue pairs and in comparative cytogenetic among distinct species. G-banding involves a treatment of chromosomes using saline solution (SSC) or diluted trypsin followed by Giemsa staining. The dark bands generated correspond to A+T rich DNA with a low concentration of genes, whereas the light bands are G+C rich DNA, presenting a high gene concentration. Due to the quality and good definition of produced bands, this methodology has been more extensively applied for studies related to mammalian chromosomes. In other vertebrates, for example avian chromosomes, the quality of the results is poor and for fish, amphibians, reptiles and invertebrates only faint bands or low traces of longitudinal differentiation have been observed. Nonetheless, there is a number of works in the literature in which this methodology has been successfully performed for analysis of these above mentioned groups. Some ideas to explain the better longitudinal differentiation of chromosomes in mammals and, to a lesser extent in birds, in comparison to other animals have been proposed. One of these explanations relates the presence of differentiated bands with genomic compartmentalization in isochores, differing in base composition, that are absent in lower vertebrates and invertebrates (Sumner 2003).

Some examples of the use of G-banded chromosomes are available in the current cytogenetic literature and also in less recent publications on mammal cytogenetics. These analyses have been used mainly for the determination of genetic homology of chromosomal segments among distinct populations or species. Moreover, it is a valuable tool for detection of chromosomal rearrangements, besides the determination of ancestral karyotypes by comparative analysis, those are useful in the studying of chromosomal evolution. Although the G-banding remains in use for mammals, other modern cytogenetic methodologies, such as chromosome painting have been more extensively used for large scale chromosomal analysis, mainly in the elucidation of primitive karyotypes, confirming and refuting some ideas proposed by G-banded karyotypes (see chapter VII). In fact, the G-banding is still in use for some aspects of chromosomal studies, being an essential technique for chromosome distinction before and after chromosomal paints and in the elucidation of inversion rearrangements, generally not revealed by chromosomal paints.

Besides the use of C-, G- and silver nitrate banding methods, the animal chromosomes have been also studied, in the lesser extent, under the focus, for example, of fluorochrome banding with chromomycin A_3 (CMA_3) (Figure 1.8i, 1.9e) and 4'-6-diamindino-2-phenylindole (DAPI) (Figure 1.9f), incorporation of bases analogous, such as bromodeoxyuridine (BrdU) and R-banding (Reverse-banding). The R-banding, similar to G-banding, can be used to demonstrate the euchromatic bands, generating a reverse pattern obtained through G-banding method. The advantage of this method is related to the possibility of easily determining the chromosomal ends, due to the usual staining of these regions. Its permits the recognition of sites of chromosome breaks more accurately (Sumner 2003). Although advantageous, this methodology has not been applied routinely for the chromosomal identification, being the G-banding more used for this subject. The BrdU incorporation method is based in the non-simultaneous replication of chromosomal DNA along the S phase of the cell cycle. The banding generated by this methodology is named "replication banding" and the generation of this pattern consists in the differential incorporation of BrdU in the DNA, replacing the thymidine nucleotides. In general, this pattern is obtained in cell culture, but the injection of this basis analogous can be also used. If the BrdU is present in the early S phase, in which the GC rich regions are replicated, the pattern obtained with BrdU is similar to R-banding, while if this basis analogous is incorporated in the late S phase, the regions rich in AT are more stained, corresponding to the G-banding. This method has been an encountered application for chromosomal studies in animal groups in which the G- and R-banding are not produced on their chromosomes, such as fish, amphibians and reptiles, although it was also used for mammal chromosomal analysis.

The fluorochromes are also useful to produce chromosomal banding, although the methodologies that use these molecules has not generally been used in the routine cytogenetics to produce differentiation of euchromatic bands, being these patterns are obtained more frequently by G- or R-banding as described above. In the historic context of cytogenetics, the use of fluorochromes was important, being that the quinacrine mustard fluorochrome staining was the first modern banding technique used (Sumner 1990). In fact, nowadays, the fluorochrome staining, mainly with CMA_3 and DAPI has been used to discriminate the base composition of heterochromatin among animals (Figure 1.9e, f). The CMA_3 shows specificity for GC base pairs and the DAPI presents affinity for the AT rich regions. Besides the use of fluorochromes in heterochromatin content studies, the CMA_3 has been extensively used to determine the number and location of rDNA regions independent of their activity, that are generally CG rich, although there are, in the literature, some examples of no correspondence between the GC bands and rDNA.

Although the animal cytogenetics have evolved rapidly and a huge quantity of information has been generated after the applicability of *in situ* hybridization (see below) on chromosomes, the basic information regarding chromosomal number and shape, together with simple banding techniques are far from becoming obsolete. These methodologies remain useful for basic and applied studies, for comparative cytogenetic, cytotaxonomy and studies of chromosomal structure and evolution. In fact, the information presented along this chapter represents only a few parts of the chromosomal studies conducted through the classical techniques in animals. It was not our objective to explore in detail all available data in the literature for this subject, but only demonstrate the importance and value of classical methods. The new technologies that will be demonstrated along this book have confirmed or refuted some hypothesis previously pointed by classical analysis. In a broad sense, the basic analysis of chromosomes continuous to contribute to the development of cytogenetics, increasing our knowledge about chromosomes and genome structure of animals.

Chapter 2

IN THE ANIMAL GENOMES

During the last two decades, scientists around the world have been using DNA sequencing techniques to decipher the genetic constitution of many different organisms. The major outcomes derived from comparative investigations on related organisms showed that most biological functions are genetically conserved within and between investigated species. Once sequenced, the DNA has provided scientists important clues about the genes and proteins that are required to generate and maintain organismal diversity.

Animals sequenced initially were chosen based on the economic orthe genome size criteria, which was a limiting factor. Also, well-studied model organisms or those showing promise to become good models had priority. Species of domesticated farm animals such as sheep, chickens, pigs, horses and cattle, as well as companion animals, such as cats and dogs, hold diverse appealing attributes that favor them to have their DNA decoded.

Since the sequencing of *Caenorhabditis elegans*, the first multicellular genome completed (The *Caenorhabditis elegans* Sequencing Consortium 1998), the development of new technologies has dramatically decreased the complexity and cost of sequencing. Subsequently, the quantity of completely sequenced genomes is rising rapidly. A growing number of large scale genome sequencing projects has been launched reaching more than 300 surveys, specifically on animals. These surveys include completed, in assembly phase and in progress projects (NCBI database). Mammalian genome projects prevail whereas non-mammalian vertebrates are poorly represented among sequenced genomes, with most of the data still to be assembled onto chromosomes.

The term genome, when referring to animals, is usually employed only to designate the information on chromosomal DNA in the nucleus called "nuclear genome". Although these organisms contain mitochondria that have their own DNA, the genetic information stored within the mitochondrial genome in a broad sense is not usually considered part of the genome. As multicellular eukaryotic organisms of the kingdom Animalia or Metazoa, animals hold several genome copies in their whole body. These copies contain numerous genes, which confer specific characteristics of the organism and are packaged in a variable number of chromosomes.

First, animal genome studies were focused on determining genome sizes and chromosome number and composition (for review Bernardi and Bernardi 1990). Moreover, other traits such as gene order, codon usage bias, and GC-content have been compared to determine what mechanisms could have produced the greatest variety of genomes that exist today (for recent overviews, see Brown 2002, Gibson and Muse 2009, Reece 2004, Gregory 2005a). Indeed, animal genomes possess a remarkable plasticity as illustrated by highly variable DNA content and chromosome number reported across taxa (Gregory 2010, Völker et al. 2010).

2.1. Chromosome Records in Animals

While some species exhibit remarkably streamlined chromosomes, others possess massive amounts of non-coding DNA (Jaillon et al. 2004, Biémont and Vieira 2006). To the best of our knowledge, the recorded chromosome number in animals ranges from 2n=2 in the nematode *Parascaris univalens* and in the jumper ant *Myrmecia pilosula* (haploid male ant have only 1 chromosome), up to 2n=372 chromosomes in the shortnose sturgeon *Acipenser brevirostrum*. However, even close- related species have been found to exhibit huge variation in chromosome number. This was clearly observed between the Chinese muntjac deer, *Muntiacus reevesi*, with 2n=46 chromosomes and the congenera Indian muntjac, *Muntiacus muntjak* with merely 6 (female) and 7 (male) chromosomes. Similarly, changes in chromosome number can be related to the occurrence of differentiated chromosome types (e.g., supernumerary or B chromosomes, discussed later in the Chapter VIII). In addition, chromatin reduction through loss of heterochromatin regions was reported in several worm species (Tobler et al. 1992). Even an extreme change in the karyotype, such as an entire chromosome elimination, has been observed in crustaceans and insects (Legrand et al. 1987, Goday and Esteban 2001), hagfish (Nakai et al. 1995, Kojima et al. 2010) and amphibians of the genus *Rana* (Ohtani 1993).

In addition to the variation in the number of chromosomes in a set, animal cells can carry multiple copies of chromosome sets, such as diploid, triploid, tetraploid and so on. The majority of animal cells have two sets of chromosomes (i.e., are diploids). Polyploid cells are larger than diploid ones, not surprising in view of the enlarged number of chromosomes in their nucleus. Although polyploidy is much rarer in animals than in plants, several examples have been recognized from lower forms such as flatworms and annelids (Gregory and Mable 2005) to more complex organisms such as reptiles (triploidy), fish (tetraploidy in salmon and octaploidy in sturgeon) and amphibians (dodecaploidy in frogs), among others (Solar et al. 1992, Birkham and Hanks 2009). Interestingly, recent findings in genome research indicate that many species that are currently diploid, including humans, were derived from polyploid ancestors (Van de Peer and Meyer 2005). These species, referred to as paleopolyploids, which have experienced ancient genome duplications and then genome reduction, are good examples of the high chromosomal dynamism into animal genomes.

In fact, there are both advantages and disadvantages of having multiple sets of chromosomes. Advantageous examples include heterozygozity maintenance, which prevents the accumulation of recessive mutations in the genomes of later generations (more common in plants than animals); gene redundancy, which confers the ability to diversify gene function over time, leading to new opportunities in evolutionary selection (Adams and Wendel 2005); the disruption of certain self-incompatibility systems, thereby allowing self-fertilization and favoring the onset of asexual reproduction. This switch in reproductive strategies may improve fitness in static environments (Woodhouse et al. 2009). On the other hand, there are innumerous disadvantages of polyploidy. These disadvantages include the relative changes between the size of the genome and the volume of the cell, leading to an increase in the surface area of the nuclear envelope (Melaragno *et al.* 1993). This event can disrupt both the correct interactions between the chromosomes and nuclear components and the desirable peripheral positioning of telomeric and centromeric heterochromatin (Fransz *et al.* 2002). Other disadvantages comprise the abnormal completion of mitosis and meiosis, which can lead to the chaotic segregation of chromatids and to the production of aneuploid cells in animals. Moreover, polyploidy can lead to dramatic changes in epigenetic regulation, altering DNA methylation and gene expression (Woodhouse et al. 2009). Despite the apparently disadvantageous consequences of polyploidy for the "standard" genome functionality, many of the polyploid organisms are fit and well-adapted to their environments.

Another noteworthy issue is that polyploidy has been linked to speciation. For instance, the first polyploid mammal, the Red Vizcacha rat, *Tympanoctomis barreraea*, a tetraploid (4n=102) rodent has been reported bearing liver and sperm cells larger than those of its diploid relatives. This species may have arisen when an ancestor, possibly the Mountain Vizcacha Rat, *Octomys mimax*, (2n=56) doubled its chromosome number, presumably by errors in mitosis or meiosis within the animal's reproductive organs (Gallardo et al. 2006). Normal mammalian sperm heads contain about 3 pg of DNA while the spermatic cells of *T. barreraea* contain 9.2 pg of DNA (Gallardo et al. 1999). However, there is a controversy over the ploidy level of *T. barreraea*. Some authors advocate that the Red Vizcacha rat actually is diploid (2n=102) and claim that the accumulation of repetitive sequences better explains its large genome size (Svartman et al. 2004a, Ferguson-Smith and Trifonov 2007).

Currently, it remains unknown as to what governs the optimal chromosome number, or the ideal DNA content in determining the whole genome size for any given species. If there is a rule for explaining the genome dynamics, certainly it is driven by a complex set of interactions among environmental, cellular and lineage-specific evolutionary factors.

2.2. GENOME FEATURES

Animal genomes differ considerably in size, ranging from 20 million base pairs in the plant-parasitic nematode *Pratylenchus coffeae* to over 130 billion basepairs in the marbled lungfish *Protopterus aethiopicus* (Gregory 2010). Moreover, significant variation was detected even between closely related species (Boulesteix et al. 2006). While most of the vertebrates studied so far have genomes bigger than 1 Gb, large amount of invertebrate species seem to contain very small genomes. However, this idea was originally sustained on the genome size acquired from model invertebrate organisms sequenced, which were chosen due to the convenience of handling their compact genomes. While the primary vertebrate species of interest (e.g., zebrafish=1.5 Gb; frogs=1.5 Gb; mice=2.5 Gb, and humans 3.0 Gb) possess typically -sized genomes for their taxa, the first two invertebrate species sequenced, *Caenorhabditis elegans* (100 Mb) and *Drosophila melanogaster* (175 Mb) have remarkably small genomes even when compared respectively to other roundworms and flies. For example, *Drosophila virilis* possess a 400 Mb genome, two-fold bigger than *D. melanogaster* genome; in the same way the fly, *Musca domestica* and the mosquito, *Aedes aegipt* have a genome with more than 940 Mb, which is larger than the genomes of the fishes fugu and medaka. Thus, the incorrect idea that small genomes are a fundamental property of invertebrates was widespread. Nonetheless, invertebrates can hold large and non-coding rich genomes, as do mammals (Peterson et al. 2009).

Animal groups can present a vast diversity of genome sizes (Gregory 2010). The comparison of distinctly-sized animal genomes has given support to numerous inferences about genome trends. In zebrafish, whose genome is about half the size of most mammalian genomes (Driever et al. 1994, Gregory 2010), initial comparisons with mammalian gene maps revealed extensive conservation of syntenic chromosome regions among vertebrates (Postlethwait et al. 1998, Amores et al 1998, Gates et al. 1999, Barbazuk et al 2000, Woods et al. 2005). Birds, for instance, have small and conserved-sized genomes due to a lower number of interspersed repeats, segmental duplications, and pseudogenes, with a narrow variation interval being the smallest genomes observed in vertebrates (The International Chicken Genome Sequencing Consortium 2004, Gregory 2010). Moreover, the evolutionary conservation of avian genomes in terms of size, overall karyotype structure (the 2n=80 has remained constant for most species), is contrary to the non-coding rich genomes of mammals (Ferguson-Smith and Trifonov 2007). Birds usually bear few macro- and many small microchromosomes and limited structural variation (Rodionov 1997, Griffin et al. 2007). Another interesting example is the duck-billed platypus – a mixture of bird, reptile and mammal – whose genome carries a highly unusual number of 10 sex chromosomes and is relatively large (comprehend about two-thirds the size of the human genome) with 50% composed of repetitive elements (Warren et al. 2008).

Strong positive correlation was found between genome size and repetitive DNA content (Gregory 2005a, Lynch and Conery 2003, Kidwell 2002). This is because the expansion of repeat families or the removal of repetitive DNA, directly alter the genome size (Lynch and Conery 2003, Petrov et al. 2000). Furthermore, coding DNA undertakes significantly less variation in size than intronic and intergenic DNA domains (Gregory 2005a). By contrast, there is a lack of correlation between the total genome size in base pairs and the number of genes identified, as well as between these identified genes and the organismal complexity. For instance, *Trichomonas vaginalis*, the protozoan that causes trichomoniasis possesses around 60,000 genes. This represents twice as many genes as in the mammalian genome, which stands for the most complex group. This puzzle circumstance known as C-value enigma (Gregory 2005a) commonly refers to genes. However the amount of non-coding DNA within the genomes also varies significantly across taxa. Distinct types of non-coding DNA and in irregular proportions come and are spread and/or lost from genomes over time and most of their effects or functions remain unresolved from chromosome to organism level.

It is already known that diverse noncoding DNA sequences may contain information that guides chromosome structural organization, centromere function and homolog recognition in meiosis (Subirana and Messeguer 2010). Also, highly conserved noncoding DNA sequences can rule out binding sites for transcription factors and, thus, interfere in the expression levels of several groups of genes (McDaniell et al. 2010, Kasowski et al. 2010). Usually, the fraction of these highly conserved noncoding bases lying outside of exons of protein-coding genes, grows with the complexity of the investigated group (Siepel et al. 2005).

Additionally, whole-genome comparisons revealed an abundance of highly conserved noncoding elements clustered along chromosomes as syntenic blocks (Zdobnov et al. 2002; Pevzner and Tesler 2003). Synteny data combined with DNA sequence information has been useful for detecting genome duplications (Myers 2008). As a classical example, the Hox genes, key determinants for the animal body plan, have a well-conserved order in distinct clusters all over the animal kingdom (Lutz et al. 1996; Santini et al. 2003, Lee et al. 2006).

Therefore, regardless of the observed plasticity in the animal genomes, (e.g., large variation in genome size and chromosome number), many genes and large portions of highly conserved noncoding elements occur in synteny on chromosomes, revealing that animal genomes are under sharp evolutionary constraints.

2.3. DRIVING FORCES UPON GENOME EVOLUTION

The raw material for evolution is represented by pre-existing DNA molecules and genes: in theory, there are no natural mechanisms creating new DNA sequences. Thus, new DNA sequences and genes come up by means of modification of earlier ones. These innovations can emerge through several ways such as intragenic mutations, transposable elements, lateral transfer, exon shuffling (including gene fusion/fission), *de novo* origination and gene duplications (Chandrasekaran and Betrán 2008). Also, deletion of genes and of other DNA fragments efficiently alters genome constitution. However, it is important to consider that such molecular mechanisms must occur in the germ line to be inherited by the next generation and afterwards, fixed or extinct. A brief description of the most effective mechanisms that alter animal genomes is discussed in the next topics.

2.3.1. Transposable Elements

Transposable elements (TEs) are DNA segments able to copy or move themselves within a genome without coding functions. Insertions of DNA transposons and retrotransposons, the two known TEs types, originate mutations and enlarge genome size. An important feature of these elements is the polymorphisms generated as a consequence of the insertion and the variability in the number of copies that can arise within and between species (Feschotte and Pritham 2007, Lankenau and Volff 2009). This addition of copies in the genome can cause mutations that can often be favorable and ultimately help to increase diversity in nature, driving the evolutionary force. Transposable elements are also important for the maintenance of the structure and the organization of chromosomes and may induce the formation of chromosomal rearrangements or may act to prevent telomere loss (Maside et al. 2001). They are responsible for important events in the development of some organisms (genetic and epigenetic regulation), changing the idea that they are just parasites present in the genome (Böhne et al. 2008). Unexpectedly, TE encoding cellular proteins (called domesticated proteins) are involved in the functioning of the vertebrate immune system (Fechotte and Pritham 2007). In insects, TE participate in transcription, chromatin remodeling and protein-protein interactions (Casola et al. 2007). A better overview of TE will be presented in Chapter V.

2.3.2. Lateral Gene Transfer

Lateral transfer refers to the origin of genes from an unrelated genome, rather than from parent to offspring. This mechanism has been observed in plants, bacteria and cellular organelles but still remains largely unknown. Some examples of lateral transfer were described in the genome of fruit fly species, by acquisition of genes of the intracellular endosymbiont *Wolbachia pipentis* (Hotopp et al. 2007), as well as in the ancestral of tunicate *Ciona intestinalis*, probably due to gene transference between a chordate and a likely extinct protostome (Syvanen and Ducore 2010). Recent studies using *in silico* approaches suggest that the incidence of lateral transfer has been underestimated (Choi and Kim 2007; Popstova and Gogarten 2007). Clearly, the majority of biological consequences of lateral transfer in animal genomes need to be better investigated.

2.3.3. Exon Shuffling

Distinct exons, either within a gene or between two nonallelic genes, are occasionally mixed in a process termed exon shuffling (Gilbert et al. 1978, Li et al. 2001). This mechanism is facilitated by the intron-exon organizational pattern extant in genomes that constantly endure recombination events. Exon shuffling usually involves both gene fission and fusion routes. Some of the combined exons will create chimeric proteins partially carrying domains or segments which are similar or identical to domains of other proteins, and usually with innovative functions. This process is referred to as neofunctionalization of the new gene and has allowed the expansion of the protein isoforms arsenal of multigene families (Patthy 1999). Many genes sharing exons were detected producing enzymes involved in the blood coagulation and fibrinolisis processes on vertebrates (Kolkman and Stemmer 2001). Another example is the primate specific chimeric gene, the *SETMAR* gene, which was created by merging an existing host gene with a transposon element (Cordaux et al. 2006)

2.3.4. Pseudogenes and *De Novo* Originated Genes

De novo originated genes found in animal genomes are often derived from noncoding DNA. For example, numerous genes with no homologues in any other species have been described in *Drosophila* (Begun et al. 2007, Levine et al. 2006). These new genes include both protein-coding and non-coding genes (i.e., pseudogenes).

The investigation of pseudogenes, originally defined as sequences that resemble known genes but cannot produce a functional protein, has revealed that many sequences once believed to be "inactive genes" are, in fact, functional. Recently, Kondo et al. (2010) evidenced that a RNA called *polished-rice* (*pri*), previously considered as long, non-coding, or "junk" RNA, which, in fact, encodes tiny peptides that control gene expression during development in the fruit fly *Drosophila melanogaster*. This and other recent findings exemplify how far the scientific knowledge is from the complete understanding of animal genomes processes.

2.3.5. Gene and Genome Duplications

Among several elements that play a role in shaping the genome, duplication events are of major importance. Distinct mechanisms are known to generate duplications, including unequal crossing over, retrotransposition, aneuploidy and polyploidy. Duplications may range from extension of short tandem repeats, individual exons, to duplication of a cluster of genes, and all the way to duplications of entire chromosomes or even entire genomes (Betrán and Long 2002).

Gene duplications are believed to have provided conditions for main evolutionary transitions towards higher complexity, including evolution of multicellularity, bilateral symmetry and vertebrate evolution (Sidow 1996, Holland 1999, Lundin 1999, Patel and Prince 2000). If this previous statement is true, there would be expected higher gene duplication rates in vertebrates than in 'lower' invertebrates. In fact, flies undergo an estimated gene duplication rate of 17 genes/million years (Hahn et al. 2007a), whereas in humans the rates are about 100 genes/million years (Hahn et al. 2007b). In the same way, Demuth et al. (2006) showed that the variation in gene duplication rates between humans and chimpanzees is estimated to be 6% higher in the former organism. It could be argued that intergenic mutations rather than gene duplications would have major duty in creating gene diversity. However, the comparison of single nucleotide polymorphisms (SNPs) between orthologous sequences of humans and chimpanzees found an estimated value of only 1.5% of divergence, refuting this idea. Based on the above mentioned data, gene duplications still seem to be more active elements than mutational events during genome evolution.

Nowadays, much the evolutionary mechanisms underpinning genome complexity have been deciphered by comparative analysis of entire genomes. However, fish for example, have naturally duplicated genomes, challenging the true assembly and annotation of a representative genome sequence. Like this, cytogenetic data works as anchors that can be integrated with accumulating genomic data (Freeman et al 2007).

For instance, Dehal and Boore (2005) gathered whole-genome sequences and physical mapping to prove that large-genome duplications were fundamental for creating genetic novelty in vertebrates. They plotted the genomic map positions of only the subset of paralogous genes that were duplicated prior to the fish–tetrapod split, showing that their global physical organization provides unmistakable evidence of two distinct genome duplication events early in vertebrate evolution. This example clearly shows the utility of merging genome sequencing and physical mapping to trace back DNA segments to their original chromosomes. Although a genome map is less detailed than the order of every DNA base (supplied by whole-genome sequencing), it allows the identification of highly informative chromosome landmarks for evolutionary studies.

2.4. CONCLUDING REMARKS

As presented, a large number of molecular mechanisms generate the genome diversity and variability thus far observed. New life forms constantly emerge in nature and are retained or lost according to the presence or the absence of selective constraints. The environment contributes in selecting a well-fit organism and in some instances, by triggering cellular processes which will lead to genome changes, most of which increases its variability and diversity. The current life frame comes from billions of years of evolutionary experiences and it is a mistake to believe that the actual status stands for the final result. Time keeps pushing out animal genomes by modeling the genetic variability and diversity which contribute decisively to life maintenance.

Less than 25 years ago, molecular biologists envisioned animal genomes as stable structures carrying the genetic birthright. Nonetheless, outstanding advances in the last few years have shown the intense dynamism experienced by DNA sequences. It is notable that the genome plasticity in response to cellular environmental changes allow a constant origin of essential novelties in the course of organisms life. Finally, genomes must be visualized as entities governed by multifactorial elements rather than by standard Mendelian inheritance patterns, which maintain the distinct DNA classes of sequences carried on chromosomes in permanent transformation.

Chapter 3

CYTOGENETICS MEETS GENOMICS AS A NEW AREA OF BIOLOGY: CYTOGENOMICS

During the first half of the 20th century, the human chromosome number was in the focus of several scientific papers, but the results obtained through tissue sections were imprecise and gave incorrect information (reviewed in Capanna 2000). The correct diploid number of human cells was only established as 2n=46 in 1956 (Tjio and Levan 1956). In later years, new techniques have allowed for the obtention of high quality chromosomes, using cell suspension preparation and cell culture. In the second half of the 20th century, several methods of chromosome staining and banding allowed for expressive progress in the cytogenetic area. After 1980, with the advances in molecular biology and genomics, cytogenetics experienced a synergy with molecular biology, allowing significant advances in understanding genomes throughout the chromosomes. Nowadays, the advances in microscopy, the application of bioinformatics, and the integration of chromosome analysis and genomic data represent promising tools for the future of cytogenetics.

The availability in the last decade of hundreds of completely sequenced eukaryotic genomes opens new avenues for the cytogenetics, with more perspectives for physical chromosomal mapping of genes and comparative cytogenetics. Although the integration of cytogenetics and genomics seems to be a recent event, the chromosome took its place of importance in the emerging of genomics. The "genome" terminology appears for the first time in the book of Hans Winkler (Winkler 1920), with reference to the "haploid chromosome set." In this way, the chromosome set represented the base for the foundation of the new area that emerges as "genomics." The connection between genome and chromosome has been often conducted in an indirect way, using linkage mapping that is based on the determination of gene positions in chromosomes according to recombination frequencies among them. More recently, the complete nucleotide sequence of several genomes also allows the obtention of chromosome maps for some species. In all of these cases, the "chromosome map" comes from indirect methods that do not involve "physical chromosome" analysis. Despite the intensive effort of studies using new genomic analyzer tools and bioinformatics, many of the genomes reported completely sequenced, still-present portions that remain as gaps due to the difficulty in correctly identifying the position and array of particular types of nucleotide sequences, such as the repetitive DNAs. In this way, the molecular cytogenetics comes to integrate nucleotide sequences with physical chromosome mapping of DNA sequences or genes. The cytogenetic can provide true "physical chromosome maps" that are of great value to anchor and support other genetic maps, such as linkage maps, restriction maps and nucleotide sequence maps.

The advances in bioinformatics permit the karyotype reconstructions of species based on *in silico* analysis of complete nucleotide sequences of genomes. The ancestral syntenies of nucleotide sequences across different groups can be established based on sequence orthologies among species. Such an approach allows the application of electronic chromosome painting (E-painting) and the foundation of "*in silico* cytogenetics" as a new perspective for analyzing chromosomes and karyotypes. Kohn and co-workers (Kohn et al. 2006) have applied *in silico* cytogenetics to a large data set of genes of humans, chickens, zebrafish and pufferfish, advancing in the reconstruction of the ancestral vertebrate protokaryotype comprising 11 protochromosomes. *In silico* cytogenetics permitted the identification of conserved linkage groups between very distant related animal groups. Ancient eumetazoan chromosomes have been recovered by comparing the human and the sea anemone (Cnidaria) genomes (Putnam et al. 2007). Conserved synteny between human and sea anemone chromosomes were identified in 40 large homologous segments, despite 700 million years of divergence among them (Putnam et al. 2007).

On the other hand, some conflicts between *in silico* (bioinformatics) and cytogenetics analyses are apparent. This can reflect the genomic sampling, limited to a few species associated with specific algorithms applied by bioinformatic tools. The increasing taxa sampling and the development of more sophisticated bioinformatic tools will allow the match between the cytogenetic and bioinformatic models (Froenicke et al. 2006). The more precise integration of cytogenetics and bioinformatics, plus the inclusion of different genetic and genomic data, will allow for a reliable reconstruction of karyotypes and the evolutionary history of groups. The integration of genomic approaches to cytogenetics signalizes to the emerging of a new area in biology, the "cytogenomics".

Chapter 4

APPLIED TECHNOLOGIES IN THE PHYSICAL MAPPING OF ANIMAL CHROMOSOMES

For several decades, the cytogenetic analysis were performed using classical karyotyping methods, which allowed a gross description of chromosome structure and organization, revealing chromosomal numbers and morphology, and sex chromosome systems in animals and plants. Even with the advance on chromosomal banding techniques, such as C-banding, silver nitrate staining, G-banding and others, the chromosomal studies remained depending on the description of banding patterns along the length of each chromosome with limited resolution. The major advance in cytogenetics has come in the last two decades with the application of *in situ* detection of DNA sequences in the chromosomes, using specific DNA segments as probes. This technique defined the transition from the classical cytogenetic era to the molecular cytogenetic era, allowing more detailed studies in the cytogenetic field and enabling the integration of molecular information of DNA sequence to their physical location along chromosomes and genomes (Schwarzacher 2003, Jiang and Gill 2006). The molecular cytogenetics has gained a great importance in chromosomal studies by allowing the direct and physical location of a specific DNA segment in the chromosomes. The molecular cytogenetics can be applied in the detection of unique or repetitive sequences, specific chromosomal regions, entire chromosomes and even whole genomes. Such an approach allows the integration of cytogenetic maps to linkage maps, elucidation of chromosome structure, and genome organization and evolution.

The *in situ* hybridization technique is based on the capacity of denaturation and specific annealing of complementary DNA chains (adenine anneals with thymine and cytosine with guanine) through hydrogen bonds between the bases attached in the DNA sugar-phosphate backbone (Swinger and Tucker 1996, Wilkinson 1999). This technique depends on a labeled probe (sequence of interest) and chromosome spreads obtaining, denaturation of the probe and chromosomal DNA (target), hybridization (probe-target DNA annealing) and probe detection (Figure 4.1).

Since the first use of *in situ* detection of a DNA sequence using a complementary DNA as a probe in cytological preparations by Pardue and Gall (1969) in the toad *Xenopus laevis*, this technique has undergone modifications related to the hybridization process, probe labeling and detection, and analysis methods increasing sensitivity, specificity and the resolution of results. Initially, the probes for *in situ* hybridization were labeled directly by radioactive isotopes ^{32}P, ^{125}I, ^{3}H and ^{35}S, but since the beginning of the 1980's, the probes started being labeled by non-radioactive molecules. Although several methods based on enzymatic reaction using alkaline phosphatase, beta-galactosidade or horseradish peroxidade were available, the most applied method in the subsequent years was based in the utilization of fluorescent elements, therefore the technique was named fluorescent *in situ* hybridization (FISH) (Pardue and Gall 1969, Forster et al. 1985, Pinkel et al. 1986, McNeil and Ried 2000, Schwarzacher and Heslop-Harrison 2000). The use of FISH permits a color era for cytogenetics and a substantial increase in the quality of the final results observed. Some information about the processes of probe obtaining, labeling and detection, and microscopic analysis for FISH experiments will be discussed along this section.

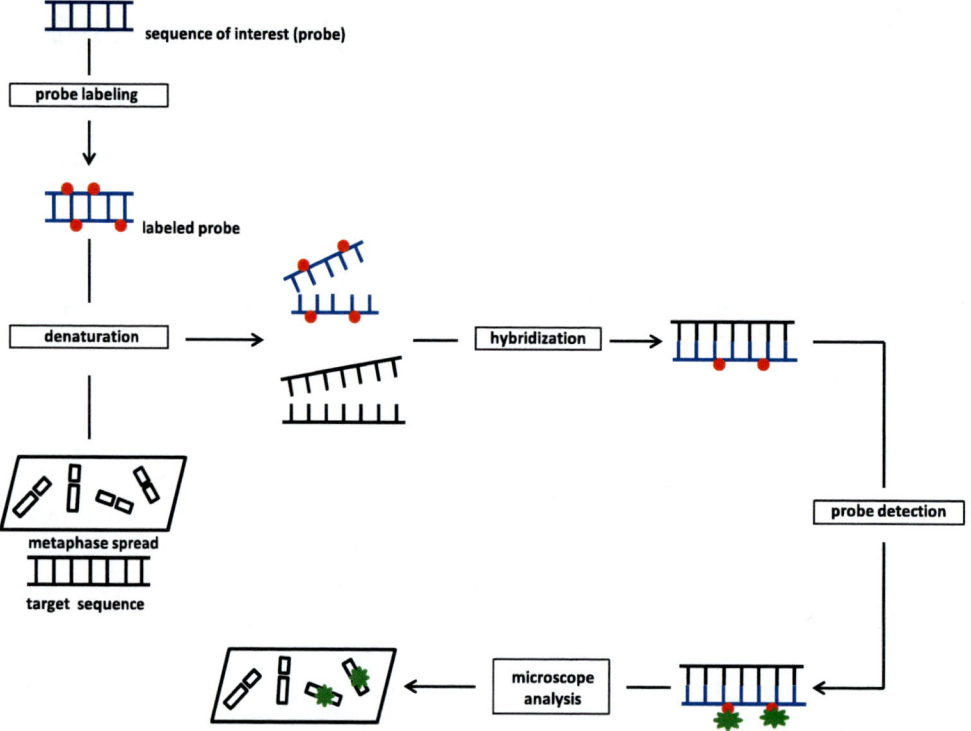

Figure 4.1. Basic steps of fluorescence *in situ* hybridization (FISH).

4.1. ROUTINE METHODS FOR PROBE OBTAINING

The advent of molecular biology, with advances in cytogenetics and cellular techniques and equipment, has permitted the generation of a range of probes to be used in chromosomal mapping. Among these techniques, the most used for this purpose are Polymerase Chain Reaction (PCR), enzymatic restriction, chromosome microdissection, flow sorting and BAC library construction (Figure 4.2).

Applied Technologies in the Physical Mapping of Animal Chromosomes 33

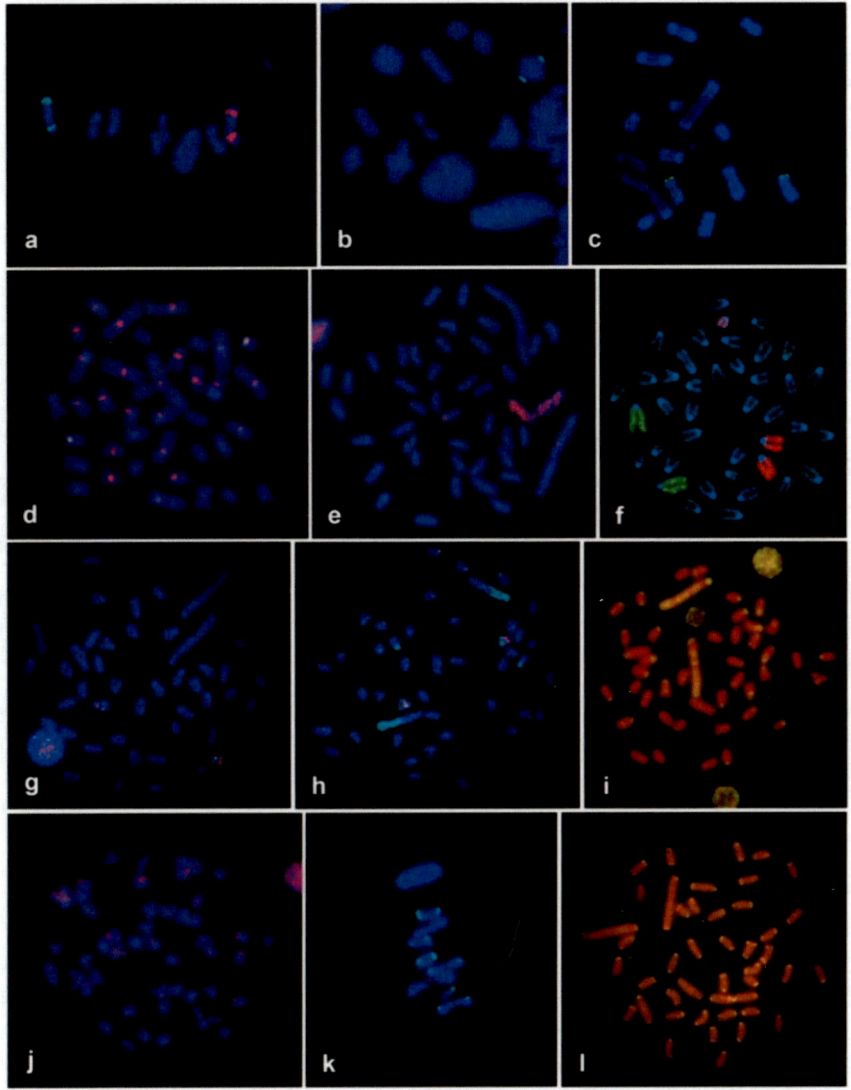

Figure 4.2. Fluorescence *in situ* hybridization results using different probes obtained by available routine methods. (a-d, j, l) probes obtained though PCR: (a) 18S rDNA (green) and 5S rDNA (red) in the beetle *Dichotomius bos*, (b) H3 histone gene in the grasshopper *Stiphra robusta*, (c) 18S rDNA in the bat *Tonatia saurophila*; (d) 5S (green) and 5S*Hind*III satellite (red) in the fish *Hoplias malabaricus*; (j) *Rex3* transposable elements in the fish *Haplochromis obliquidens*; (l) telomeric probe in *Oreochromis niloticus*; (d) satellite DNA 5S*Hind*III (red) obtained by enzymatic restriction in the fish *Hoplias malabaricus*; (e) chromosome paint probe obtained by chromosome microdissection of chromosome pair 2 of the fish *Oreochromis niloticus*; (f) chromosome paint probes obtained by flow sorting of pairs 2 (green) and 3 (red) in mouse; (g, h) BAC-FISH in *O. niloticus* using BAC clones of green opsin (green) and blue/red opsin (red). Note the differences in the green signal intensity between "g" and "h"; (g) BAC with low quantity of repetitive DNA and (h) BAC for the same chromosomal region containing high amount of repetitive DNA; (i) BAC enriched of repetitive DNAs hybridized to *O. niloticus* chromosomes; (k) FISH of C_0t-1 DNA fraction in *Dichotomius bos*; **The FISH figures "c" and "f" are courtesy of Sotero-Caio CG and Baker RJ (Texas Tech University, USA) and Yang F (Sanger Institute, England), respectively.

These techniques have permitted the generation of probes of the whole genome of one species, whole chromosomes or sub chromosomal regions, such as centromere, telomere, specific arms, specific chromosomal locus or dispersed elements, constituted by highly and moderately repetitive DNAs or single copy sequences (Figure 4.3). These probes have been used to analyze condensed chromosomes in metaphases, interphasic nucleus, cells in initial division stages and in distended DNA fibers (Fiber-FISH), that are used to study the karyotypic ongoing in distinct animal groups.

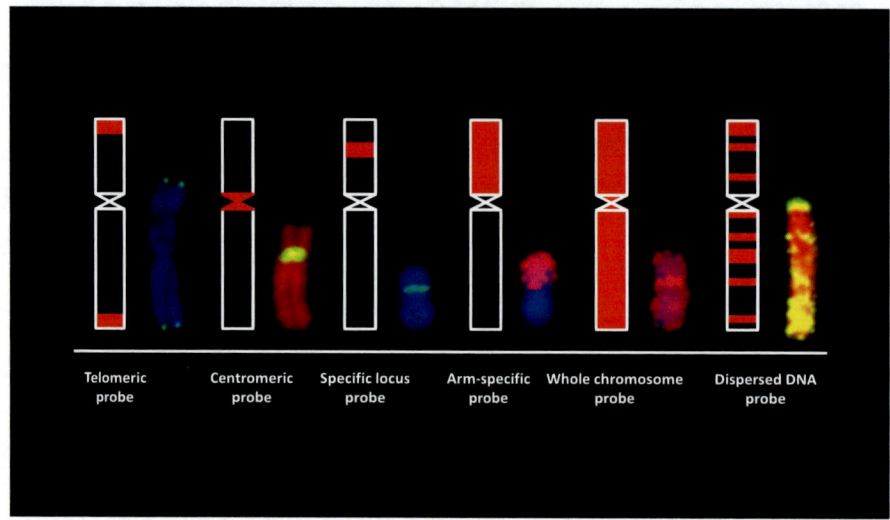

Figure 4.3. Most applied probes in fluorescence *in situ* hybridization experiments.

The most common probes used in FISH experiments are composed of repeated DNAs organized *in tandem* or dispersed in the genome. The *in tandem* repetitive DNAs most applied in FISH experiments include the satellite DNAs (Figure 4.2d), multigenic families (Figure 4.2a-c), and telomeric (Figure 4.2l) and centromeric sequences. In general these probes provide well-visible signals, due to their abundant repetition and distribution organized in large blocks along the chromosomes.

Some satellite DNAs, and the multigenic families of ribosomal DNAs (5S and 18S rRNA genes), have been isolated from distinct animal species, including invertebrates and vertebrates, providing good markers for chromosome identification. The dispersed repetitive DNAs are represented by the transposons and retrotransposons (Figure 4.2j) that are moderately or highly repeated sequences dispersed throughout the genomes, which can form viewable blocks, next to the genes, associated with a specific genomic area, such as ribosomal DNAs and heterochromatins, or dispersed in euchromatic regions of the chromosomes.

Among the methodologies for probe obtaining, the most used technique is the PCR (Figure 4.4a). Compared to the other methods, PCR represents the easiest and cheapest procedure, and even laboratories with a minimum condition in molecular biology can use this technology to obtain probes (see latter in this topic and in Table 4.1). As a consequence of these advantages, the most part of manuscripts that have been published in the specialized literature use probes obtained by PCR (see published manuscripts on specialized journals, such as Chromosoma, Chromosome Research, Cytogenetic and Genome Research, Genetica, among others). In general, the PCR is used to synthesize probes containing known repetitive

DNAs, such as ribosomal DNAs (rDNA), histone genes, transposable elements, and telomeric sequences. This technique consists of the amplification of DNA strands from the genomic DNA using a specific small oligonucleotide primer that anchors the start of DNA polymerization by a DNA polymerase enzyme. There are some examples of conserved primers designated to obtain sequences for animal genomes through PCR reaction, i.e. for 5S and 18S rDNAs, histone genes, and transposable elements (Martins and Galetti 1999, Pineau et al. 2005, Zhang et al. 2007, Teixeira et al. 2009, Cabral-de-Mello et al. 2010a). For highly conserved sequences, such as RNA transcribing genes, the primers are universal, and primers designated to insects, fishes, mammals or other groups can be used to amplify the same sequences in distantly related taxa.

Table 4.1. Advantages and disadvantages of the probe obtaining routine methods

Probe Obtention Methodology	Advantages	Disadvantages
PCR	Low cost and rapid probe obtaining in relation to the other techniques; Facility in laboratory manipulation; Isolation of known sequences that facilitates the analysis; Use of the same primer/probe in non related taxa; Direct labeling of the sequence of interest;	Isolation of only known DNAs that limits the studies;
Enzymatic Restriction	Possibility the isolation of not described repetitive sequence; Useful for studies related to centromere, telomere, heterochromatin organization, and sex and B chromosome structure;	Laborious technique: it is necessary the test of different restriction enzymes, cloning and nucleotide sequencing; More expensive in comparison to PCR;
Chromosome Microdissection	Possibility the isolation of entire chromosomes or even specific chromosome regions; More useful for non- mammalians animals;	Probes with low complexity, in general, it is not useful for cross-species painting; Difficulty for distinction of specific target chromosomes;
Chromosome Flow-Sorting	Possibility of isolation of all chromosomes from interest species; Probes with high complexity, allowing the cross-species, cross-genus and cross-order painting;	Necessity of cell culture with high amounts of metaphases; Difficulty in chromosome separation in some groups, mainly non mammals;
BAC library	Possibility of mapping of functional gene sequences; Possibility of mapping single-copy genes; Facility in the manipulation of constructed libraries; Integration of linkage and cytogenetic maps;	Dependence in the maintenance of a genomic library or a laborious work in the construction of a library from the species of interest; Small signals, that sometimes difficult microscope analysis;

In relation to less conserved sequences, such as histone genes, it is possible to design degenerated primers to be used in DNA amplification of sequences in non- related taxa. On the other hand, other sequences are more variable and specific primers need to be designated for restrict groups. That situation is common for transposable elements, which presents extensive variations among distinct taxa, caused by differences in the evolutionary dynamics of repeats with a consequent rapid sequence modification.

When the probe or primers of interest are not available, an easy and rapid way for obtaining a specific primer for less explored taxa is the use of DNA sequences deposited in the National Center for Biotechnology Information (NCBI). After searching NCBI, the retrieved sequences can be explored with programs available in the internet, such as ClustalW (www.ebi.ac.uk/clustalw) *(Thompson et al. 1994) for alignment, and primer3 (www.frodo.wi.mit.edu) (Rozen and Skaletsky 2000) for primer design*. These primers can be designed in conserved DNA regions and for more divergent DNA sequences, it is more suitable the design of degenerated primers from reverse translation of protein sequences, increasing the success of these primers in non- related groups.

The products obtained by PCR can be directly labeled (as explained later in this topic) or can be cloned, using ordinary bacterial plasmids. The cloning gives the advantage of the storage of the sequence of interest for a long time and the obtention of a large amount of probe that can be easily obtained by bacterial plasmid DNA extraction.

The enzymatic restriction consists in the cleavage of double DNA strands using restriction endonucleases that cuts the DNA in one specific nucleotide sequence (Figure 4.4b). This molecular approach can be used primarily to obtain probes of highly repetitive DNAs, such as satellite DNAs (satDNA) and transposable elements. The main difference related to the PCR is that in this technique, in general, it is isolated an unknown repetitive DNA sequence, and this sequence can be used mainly in related taxa, due the extensive variation in the profile of the repeated sequence even among related groups. This technique is more laborious and expensive than the PCR, due to the necessity of testing distinct enzymes to isolate a highly repetitive and useful sequence (Table 4.1). Moreover, the sequences obtained by enzymatic restriction need to be purified from agarose gel, cloned and sequenced to obtain specific information about the isolated element to be used in chromosomal studies. Southern blot experiments can also give good contributions in relation to the genomic organization of the isolated sequence. After the sequencing of the isolated repetitive DNA, primers can be designated to test the presence of this sequence in other taxa or to generate probes of this element for other species, facilitating the process of probe obtaining. This strategy has been mostly applied to obtain specific probes of satDNAs frequently used in studies of chromosome evolution, chromosome identification, B chromosome origin, centromere structure and sex chromosome evolution in insects (Cabrero et al. 2003a, Abdelaziz et al. 2007, Palomeque and Lorite 2008, Kuhn et al. 2008, 2009), mollusks (Biscotti et al. 2007, Petrović et al. 2009), fish (Phillips 2001, Caputo et al. 2009, Mazzuchelli and Martins 2009), and mammals (Adega et al. 2008, Matsubara et al. 2008, Acosta et al. 2009), among other groups.

Besides the isolation of repetitive DNAs using PCR and enzymatic restriction, another assay for this purpose is the use of C_0t-1 DNA, a fraction of genomic DNA elements enriched for highly and moderately repeated DNAs (Figure 4.2k). This methodology is based on the reassociation kinetic of DNA strands, that is faster for repetitive sequences than to low copy number elements (Britten and Kohne 1968). It consists basically of the denaturation (at 95°C) of fragmented genomic DNA of interest, reannealing in specific conditions (65°C, for distinct times), and the treatment with the enzyme S1 nuclease (at 37°C), that is active against single-stranded DNAs represented in this assay by low copy DNAs (Figure 4.4c) (Zwick et al. 1997). In fact, the C_0t-1 DNA fraction is used to "block" nontarget-specific DNAs and has been extensively applied in studies of chromosome painting and BAC-FISH to eliminate background. On the other hand, this methodology has shown to be a promise tool for isolation of non-specific repetitive sequences to be used as probes in animal cytogenetic, focusing in problems of B chromosome origin, heterochromatin differentiation, sex chromosomes evolution and genome organization (Ferreira and Martins 2008, Cabral-de-Mello et al. 2010a). In the same way, DOP-PCR (Degenerate Oligonucleotide Primed-PCR) that is based in the use of degenerated oligonucleotides can also be applied in the obtention of repeated sequences of the genome (Mazzuchelli and Martins 2009).

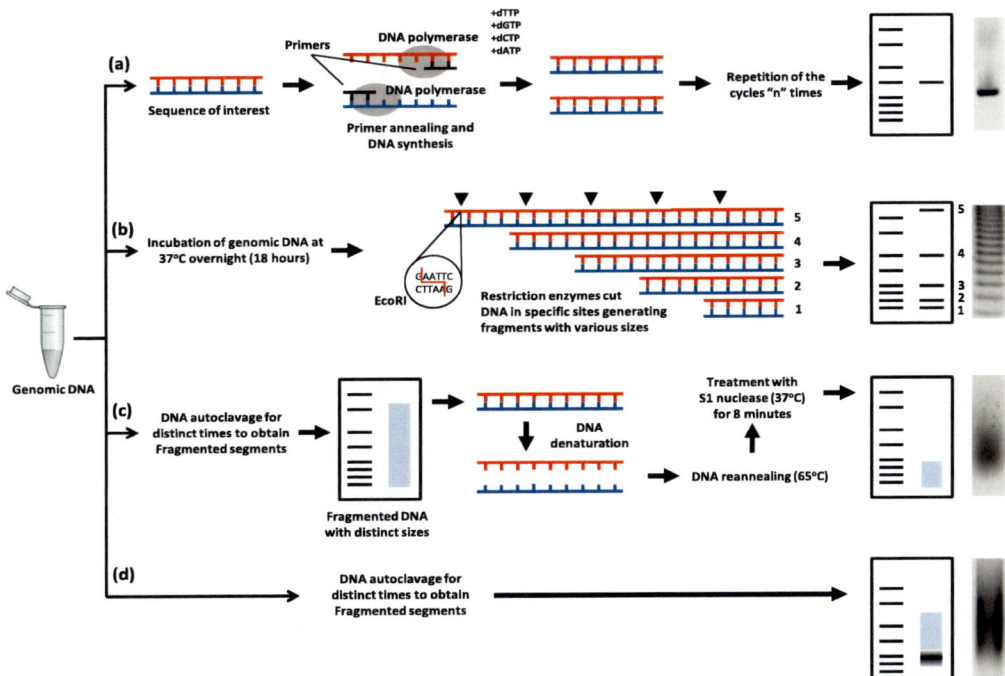

Figure 4.4. Methods of probe obtaining using as model the genomic DNA. (a) PCR; (b) Enzymatic restriction; (c) C_0t-1 DNA; (d) total genomic DNA used for GISH.

Another method used in the isolation of animal probes is the chromosome microdissection (Figure 4.5a). This methodology was first applied in polythene chromosomes of *Drosophila* (Scalenghe et al. 1981) and some modifications were introduced in the application of the technique in mammals, birds, insects, fish and other organisms. This technique permits the isolation of whole chromosomes, or specific chromosomal regions directly from the metaphases plate using micromanipulation. The technology can be useful for studies of chromosomal rearrangements, chromosome evolution, sex chromosomes and origin of B chromosomes. When this method was developed, chromosome probe hybridizations were conducted from the construction of libraries through microcloning of isolated sequences, a laborious and expensive methodology (Guan et al. 1994, Saitoh and

Ikeda 1997). Nowadays, the microdissected elements are directly amplified by PCR, frequently using a degenerate universal primer (DOP-PCR) and their products can be labeled to be used for painting (Wesley et al. 1990, Meltzer et al. 1992, Telenius et al. 1992). The most important improvement of the technique in the last years was the introduction of the inverted microscope (Senger et al. 1990), laser-based microdissector and the DOP-PCR reaction (Guan et al. 1992, Meltzer et al. 1992, Saioth and Ikeda 1997, Yang et al. 2009). These modifications allowed the decrease amount of DNA template of microdissected elements from 100-200 to 20-40 or even lesser, and higher precision in the chromosomal region to be microdissected (Yang et al. 2009). Moreover, some modifications have been done in the chromosome preparations, improving the recognition of the target chromosome, including the use of G-banded and C-banded metaphases and meiotic plates.

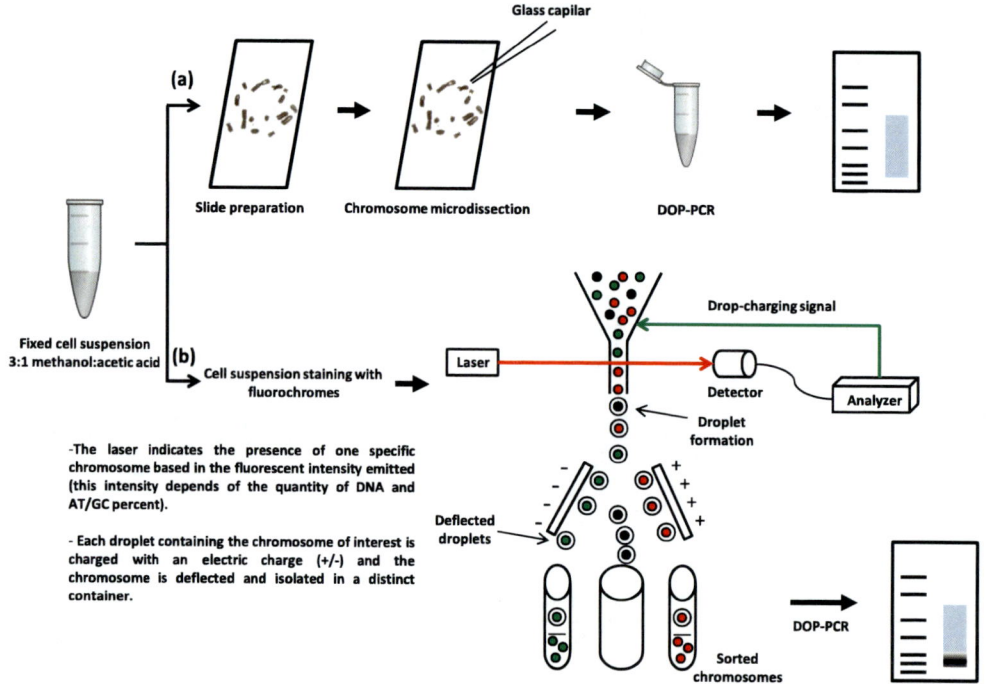

Figure 4.5. Methods of probe obtaining using as model metaphasic chromosomes. (a) Chromosome microdissection; (b) Flow sorting.

Besides the use of microdissected chromosomes to generate chromosomal paint probes, another methodology which can be used for this purpose is the flow sorting (Figure 4.5b). The flow-sorting uses the flow cytometry and sorting for separation and purification of mitotic metaphasic chromosomes (Carrano et al. 1983). This technique is based on the separation of chromosome populations by the difference in their size, morphology and DNA content that produce specific fluorescence intensity for each chromosome, generating the flow-sorted karyotype. Individual chromosomes may be isolated using a high speed liquid stream (containing the chromosomes) that is converted into droplets. The specific droplets containing the chromosomes of interest are electrically charged and are deflected electrically by a passage through an electrostatic field (Givan 2001). The use of flow-sorted chromosomes is most concentrated in studies of chromosome evolution of mammals, due the difficulty of the establishment of cell cultures and the precise separation of chromosomes in the other non mammalian groups. The difficulty of using flow-sorting in non- mammalian species is related to the symmetry in the karyotype and the chromosome compartmentalization of some species, which causes the non- precise separation of the chromosomes (Figure 4.6). On the other hand, some modifications applied in non- mammal vertebrate groups have permitted the isolation of, at least, specific chromosome groups or part of the karyotype from fish, birds and reptiles, although the results are scarce. These probes have been constantly synthesized for studies of mammalian chromosomal evolution and detection of chromosomal aberrations in cancers. Marketable probes are available for human (Telenius et al. 1992) and mouse (Rabbitts et al. 1995) karyotypes. The flow-sorting is more advantageous in relation to the microdissection generating probes with more quality and complexity, due to the efficiency in the quantity of chromosomes isolated, that permits studies using distantly related species, such as the cross-species painting, cross-genera paints even cross-order paints (Table 4.1).

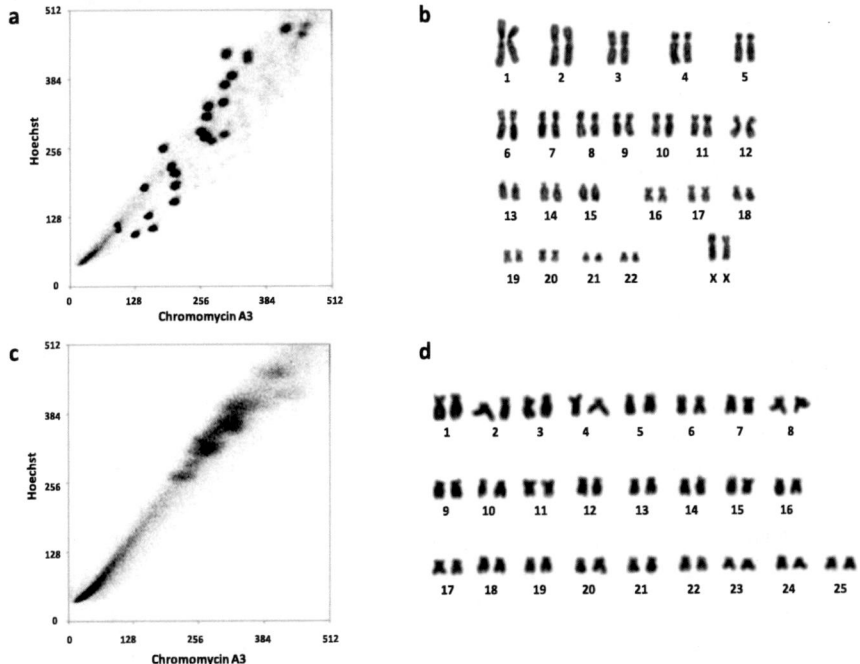

Figure 4.6. Chromosome flow sorting graphic of human female (a) and zebra fish (c) karyotypes showing the differences in chromosome separation patterns. Note the difference in the karyotype structure between humans (b) and zebrafish (d). The flow sorting graphics are courtesy of Yang F (Sanger Institute, England).

A bacterial artificial chromosome (BAC) library is a collection of cloned inserts greater than 100 kilobases (kb), usually using *Escherichia coli* (Figure 4.7). The construction of these libraries with the ability to accommodate such large inserts is advantageous and it provides an easy access to stable DNA for manipulation (Miyake and Amemiya 2004). This technology has been encountered in the application in genomic studies, such as genome sequencing, positional cloning, microsatellite and gene isolation, cDNA selection, transgenic construction and in the physical chromosome mapping through FISH, using these elements as probes (Li et al. 1999, Beck et al. 2001, Lander et al. 2001, Gong et al. 2003, Chen et al. 2004, Miyake and Amemiya 2004, Powers and Amemiya 2004, Romanov et al. 2005, Yasukochi et al. 2009).

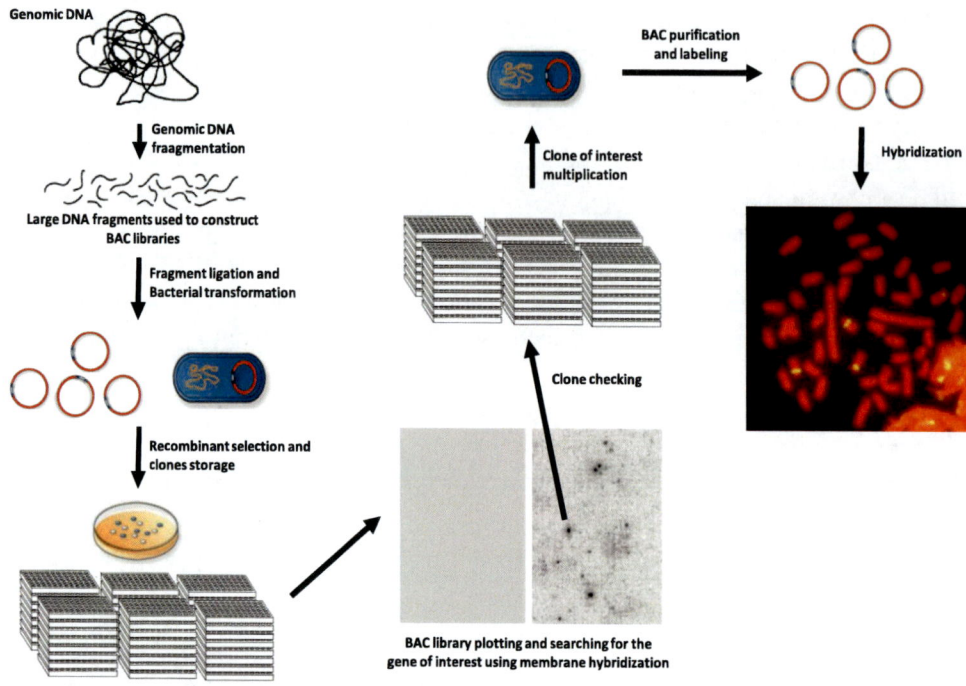

Figure 4.7. BAC-library obtaining, screening and selection of specific BAC for *in situ* hybridization.

The use of BAC inserts has permitted the integration of linkage and cytogenetic maps and has been useful in comparative cytogenetic and genomics. In general, the use of BAC-FISH is applied in studies of a specific locus region related to a gene or linkage group of interest. Moreover, some studies have been conducted using BACs with a high quantity of repetitive DNA in understanding genome organization and evolution, for example, in vertebrates (see Figure 4.2i) (Ferreira and Martins 2008, Cheng et al. 2009, Poletto et al. 2010b). Although most studies using BAC-FISH have been conducted in mammals, there is literature information concerning fishes, birds, mollusks, insects, and other groups (Aerts et al. 2003, Romanov et al. 2005, Cnaani et al. 2007, Huan et al. 2009, Ocalewicz et al. 2009, Yasukochi et al. 2009). The major difficulty in the application of BAC-FISH is the availability of BAC clones, which, in general, depends on genome sequencing projects. Nowadays, with the rapid advances in the genome sequencing technologies, the availability of BAC libraries for a range of species is increasing.

The probe obtaining methods described above are the most common used in molecular cytogenetic, although there are other approaches also in use, such as the synthetic oligonucleotides, cDNA fragments and whole genomic DNA (Schwarzacher et al. 1989, Svartman and Viana-Morgante 1999, Schwarzacher and Heslop-Harrison 2000, Matsuda et al. 2005, Srikulnath et al. 2009, Valente et al. 2009). The synthetic nucleotides can be labeled and located by ordinary FISH procedures using a specific motif with abundant repetition, i.e. the telomeric and microsatellite sequences. The oligonucleotides are relatively cheap and fast to order, and the labeling can be incorporated during the synthesis or by chemical modification currently by end labeling (Schwarzacher and Heslop-Harrison 2000). According to Schwarzacher and Heslop-Harrison (2000), in the future when *in situ* methods become more sensitive and allow the detection of single copy nucleotides sequences, the use of synthetic probes will become more common and universal. The cDNA probe obtaining method consists of the use of cloned DNA sequences obtained by RT-PCR (Reverse Transcription-PCR) technique, and permits the chromosomal mapping of functional genes or libraries of EST (Expressed Sequence Tags). This method has been recently applied with success in reptiles and birds (Matsuda et al. 2005, Srikulnath et al. 2009). The use of whole genomic DNA from a species as probe can be useful and informative for the analysis of hybrids, genome relationship, chromosomal evolution and introgression. Moreover, the genomic DNA is applicable to prevent unspecific probe hybridization in FISH experiments. Actually, the use of genomic DNA is a modification in the FISH technique called genomic *in situ* hybridization (GISH), explained in more details below.

4.2. PROBE LABELING AND DETECTION

In the FISH technique, the probe detection and results obtained are based in the observation of fluorescent molecules by means of an epifluorescence microscope. The fluorescent systems permit a better definition of hybridization signals in relation to the radioactive or enzymatic methods, facilitating the distinction between dirt, background and real signal.

The fluorescent molecules can be incorporated directly in the DNA sequence (direct labeling) or can be coupled to another small molecule that recognizes a marker molecule (hapten) bounded in the probe (indirect labeling). In the direct way of the labeling method, the probe is labeled with a nucleotide bounded to a fluorochrome, like Texas red, Cy5, among others, which is incorporated into the probe sequence instead of an ordinary nucleotide. In this method, no special immunocytochemical visualization procedure is necessary and the probe could be observed after the hybridization step, thus making this method faster and resulting in low background in the results, but slightly less sensitive, that is attributed to the presence of the low number of fluorescent molecules in each probe fragment (Schwarzacher and Heslop-Harrison 2000).

In the indirect labeling method, modified nucleotides are incorporated into the probes and a detection step is required, being that this method is more sensitive, but in the same time, with more background. Although there are a variety of haptens for label and anti-hapten antibodies, the marker molecules used more in the indirect labeling are the biotin and digoxigenin (DIG) coupled to dUTP nucleotides, which are incorporated into the probes covalently in substitution of the thymine, as described in the method of direct labeling. The biotin is an H vitamin and the digoxigenin is a steroid isolated from the foxglove plant (*Digitalis purpurea* and *D. lanata*). Both molecules are coupled in the carbon five of the pyrimidine ring by a long carbon ring that is important to avoid the reduction of the efficiency recognition of the marker molecule and the stereochemistry interference between the target DNA and the probe. After the hybridization process, the probes labeled with biotin or digoxigenin need to be detected by means of antibodies or a molecule with high affinity to the marker molecule linked with an appropriate fluorochrome. In general, it can be used to detect the biotinilated products, the avidin, a molecule with high affinity to biotin, and the digoxigenin the anti-digoxigenin antibody coupled to a fluorochrome. The initial FISH experiments were carried out using the fluorescein isothiocyanate (FITC) as a fluorochrome, but nowadays, the use of some distinct fluorochromes permits the use of distinct probes in the same metaphases offering precise and clear hybridization signals.

The most common fluorochromes used in FISH experiments bounded to the probes emit signals in red (Cy3, rhodamine, texas red) or green (FITC, Cy5) colors, and the chromosomes can be counterstained for example in red (propidium iodate) or blue (4',6-diamidino-2-phenylindole - DAPI). The chromosome counterstaining can be conducted either separate or directly mixed with the antifade solution, which prevents the brightness fluorochrome decay. Moreover, some other flurorochromes are available with other colors, and the combination in distinct equimolar proportions of the basic colors can generate probes with some distinct colors, frequently used in multi- color FISH experiments (see below).

Independent of the type of probe labeling with direct or indirect methods, the modified nucleotides can be inserted in the probes by enzymatic methods through random priming, nick translation, end labeling, *in vitro* translation (used for probes of RNA) and PCR. The nick translation method uses the activity of two distinct enzymes simultaneously, the DNAse I and *Eschericia coli* DNA polymerase I. The DNAse I in the presence of the cofactor Mg^{2+} acts as a single stranded endonuclease, creating nicks randomly in both DNA strands of the probe; and the DNA polymerase I has exonuclease and polymerase activity polymerizing the cut strand in the 5'-3' orientation and removing the nucleotides in the 5'-3'and 3'-5' orientation, with proof-reading activity. In addiction to the enzymes, in the nick-translation the non-labeled four nucleotides (dATP, dCTP, dGTP and dTTP) areused and a labeled nucleotide, frequently dUTP or dATP, will substitute part of the dTTP in the probe. This process results in DNA strands with the same genetic information but with labeled nucleotides inserted (Figure 4.8a).

The PCR and the random priming labeling methods use the same principle of polymerase action and labeled probes are produced through the synthesis of a new DNA chain. Primers are used in the reaction with non-labeled nucleotides (dATP, dCTP, dGTP and dTTP) and a labeled nucleotide, which is randomly inserted in the new DNA sequences by polymerase action, obeying the complementarities with the DNA template, resulting in labeled probes. In the PCR labeling amplification in multiple cycles of DNA denaturation, primer annealing and DNA replication using specific primers by polymerase I action, generates new labeled DNA chains (Figure 4.8b). In the case of the random priming method, a mix of degenerated primers (each one composed by six nucleotides) is used, that contain almost all combinations of A, T, C and G, and initiates the new DNA chains in random positions in the template DNA chain, so that almost every bit of templates DNA is covered by these primers (Figure 4.8b). Another difference from the PCR labeling method is that in the random priming labeling, the single-stranded DNA is synthesized using the Klenow fragment of *E. coli* DNA polymerase I, which has only the site responsible for the 5'-3' polymerase action (Figure 4.8c).

The end labeling and *in vitro* transcription are less used as labeling probe methods in FISH experiments. The former uses the enzyme terminal deoxynucletidyl transferase (TdT) and a DNA polymerase, which is responsible for the addition of nucleotides on the terminal 3'-OH in a single or double DNA strand. In the *in vitro* transcription method, a labeled RNA molecule (riboprobes) is generated using the action of a RNA polymerase that transcribes a

DNA cloned sequence. The riboprobes are frequently used to RNA targets, and in few cases, are used to hybridize against chromosomal DNA (Schwarzacher and Heslop-Harrison 2000).

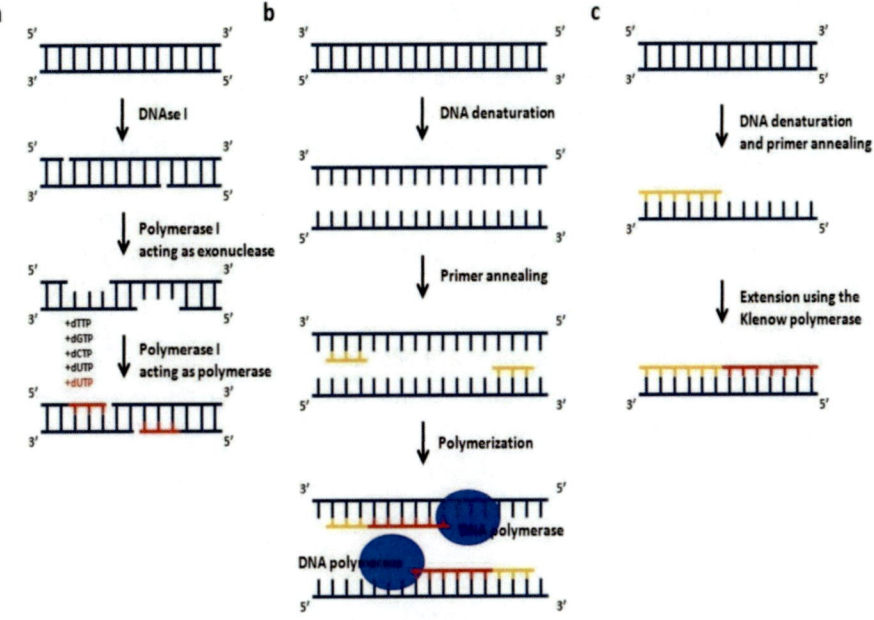

Figure 4.8. Most applied labeling methods for FISH. (a) Nick-translation; (b) PCR labeling; (c) Random priming. The primer is indicated in orange and the extended labeled DNA strand in red. For more details see the text.

The use of distinct methods of probe labeling described above basically depends on the size of initial DNA sequence and the amount of DNA in the reaction. For long DNA sequences, the nick translation labeling method is useful, that in general, give probes with lengths about 200-300 bp. The use of appropriate DNA quantity in this reaction is important because the DNA molecule is not amplified along the nick translation reaction. The random priming and PCR methods are suitable for probes using low initial DNA amounts and for shorter fragments than used in the nick translation. The size of the probe is an important parameter, due to influences in the hybridization results, where small probes have more facility of detachment from the target DNA, while the long probes can present difficulty in the cell/nucleus penetration.

4.3. RECENT ADVANCES AND APPLICATION OF FISH TECHNIQUE

During the recent years, some modifications in the traditional FISH technique were introduced, increasing the use of this methodology to solve distinct biological questions in basic research and clinical genetics. Among the advances, it can be referenced the increase of resolution in the identification of chromosomal rearrangements and the number of distinctly labeled probes that can be hybridized at the same time. Several technical modifications created to analyze human and plant chromosomes have been applied for animal chromosome analysis, such as comparative genomic hybridization array (array CGH), multicolor-FISH, fiber-FISH, and genomic *in situ* hybridization (GISH).

The GISH technique is based on the extraction of total DNA of one species (Figure 4.4d), labeling and application of this pool of DNA sequences in *in situ* hybridization experiments with cells from another related organism using cytological preparations (revised by Stace and Bailey 1999, Kato et al. 2005). This technique was first developed in animals for studies of hybrid cell-lines (Pinkel et al. 1986), and in plants, it was first used in 1987 by researches from the Plant Breeding Institute, Cambridge (Schwarzacher et al. 1989). It has been mainly used for discrimination of parental genomes in interespecific and intergenetic hybrids and allopolyploid plant species. Since the first use of this methodology, some research groups have applied this method for other purposes, such as in studies of chromosome positioning, analysis of B chromosomes, and comparative cytogenetic and genomics. Although most studies of GISH are concentrated in plants, there are few but exciting examples for use of total genomic DNA as probes in animal chromosomes, i.e. in marsupials (Svartman and Viana-Morgante 1999), rodents (Houseal et al. 1995), fishes (Valente et al. 2009), salamanders (Bi et al. 2009) and insects (Bressa et al. 2009) showing that this methodology is a promising tool in the investigation of genome structure, chromosome evolution and intergenomic exchange in animals, related to autossomal complement and sex chromosomes. This methodology does not permit the generation of precise data about organization and evolution of specific chromosomes, but it allows a gross comparison of chromosomes and genomes of related species without the isolation of specific chromosomes or chromosomal segments (Svartman and Viana-Morgante, 1999). Moreover, this methodology is inexpensive and does not require DNA cloning and sequencing.

Also related to the overall genomic analysis, the CGH array can be applied on chromosomes or in the genomic DNA. The CGH array on chromosomes was first described by Kallioniemi et al. (1992) and later it was applied essentially in studies of gains and losses of genomic parts in human clinical cytogenetic, enabling high-resolution and genome-wide screening of segmental genomic copy number variations (CNVs). The principle of this technique for humans is based on the differential labeling of a normal and a pathological DNA (i.e. extracted from cancer cells) and competitive hybridization against a normal metaphase chromosome spread, detecting gains and losses of chromosomal/genomic parts by differences in signal intensity (Forozan et al. 1997). The standard pattern in this methodology is the label of the tumor DNA and the normal DNA on equal grounds with distinct fluorochromes, mix of the labeled probes and hybridization against normal metaphases. After the hybridization, the profile of each chromosome is analyzed separately to investigate the gains and losses.

In fact, in animal studies the CGH array technique can be applied in a similar path to that applied to humans, both in chromosomes or using only the genomic DNA. Studies of descriptions of tumor conditions, genomic imbalance, chromosome aneuploidy or elimination were most frequently conducted in domestic animals and in animals with economic importance (Dunn et al. 2000, Thomas et al. 2007, 2009, Sakai et al. 2007, Hornak et al. 2009). In comparative and evolutionary studies, this technique has been recently applied in the detection and measuring of structural variations in the genome, showing inter-specific genomic differences among related species, and could also find application in phylogenetic studies, but there are, until now, few results in the literature related to chromosomal analysis

(Toder et al. 1998, Traut et al. 1999a, Griffin et al. 2008, Dávila-Rodríguez et al. 2009, Mitra et al. 2009). The use of CGH in animal chromosomes has been useful to visualize sequence homology of clusters in related species (Traut et al. 1999a), determination of species-specific chromosomal regions (Toder et al. 1998), and with most frequency, in elucidation of sex chromosome establishment in vertebrate and invertebrate groups without strong differentiated sex chromosomes, such as fish (Traut et al. 2001), crustaceans (Barzotti et al. 2000), insect groups (Traut et al. 1999b, 2001, Vítková et al. 2007), amphibians (Abramyan et al., 2009) and reptiles (Ezaz et al. 2005, 2006a), being a powerful methodology for this purpose.

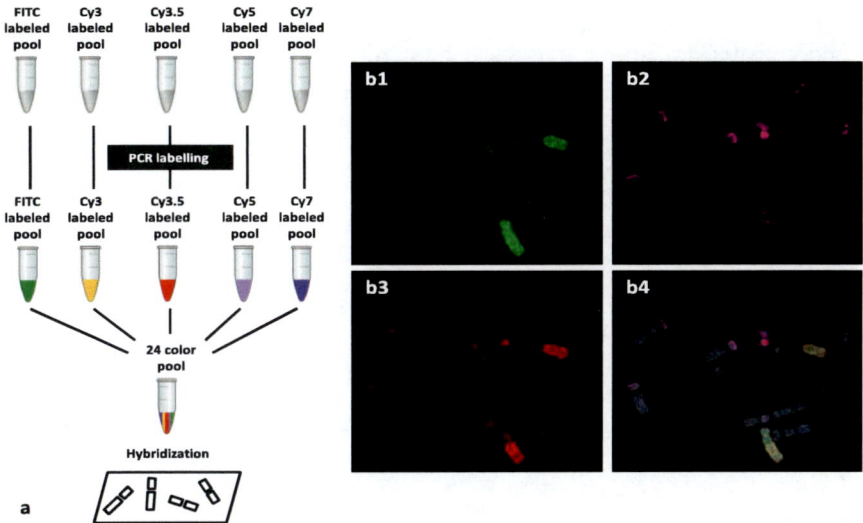

Figure 4.9. Multi-color FISH. (a) probe sets obtaining by combination of distinct fluorochromes for each chromosome; (b1-b3) sequential probe capture using the distinct probes labeled by three distinct fluorochromes for mFISH experiment in the bat *Carollia brevicauda*, (b4) overlapping of all chromosome probes in the same metaphases. In red Cy3 (Y2 chromosome probe), green FITC (X chromosome probe) and purple Cy5 (Y1 chromosome probe). Images b1-b4 are courtesy of Pieczarka JC (Universidade Federal do Pará, Brazil).

In studies about karyotype repatterning and chromosomal ongoing, the probes obtained by the flow-sorting technique are frequently used in reciprocal cross-species chromosome painting assays to establish the chromosomal homeologies among related species. These studies have been conducted mainly in mammal representatives and improvements have been conducted along the recent years. One important improvement in the whole chromosome paints (WCP) was the use of some probes obtained from distinct chromosomes and hybridized together in the same metaphase spreads, the multi-colorFISH (mFISH) (Figure 4.9b).

The mFISH assays are efficient in the description of chromosomal rearrangements and its principle was described in 1989 (Nederlof et al. 1989). Since 1996, this method has been routinely applied in the simultaneous analysis of the 24 human chromosomes in mFISH paints and spectral karyotyping (SKY). The use of some probes in the same metaphase is advantageous due to the possibility of analysis of distinct probes in the same preparation, generating economy of samples, reactives and time. Moreover, the mFISH allows an easier and more precise description of the junction points of the different syntenic blocks in comparison with results obtained by single FISH.

The generation of probes with distinct colors is based on the combination of distinct fluorochromes for the same sequence of DNA (probe). For example, using two fluorochromes the labeling of three probes with distinct colors is possible, two using the isolated fluorophores, i.e. green or red, and one with the equimolar combination of the two fluorochromes, green plus red which is yellow. The formula that determines the number of probes using combinatorial labeling assay is $C = 2^N-1$, where C is the number of different colors that can be distinguished when N different fluorochromes are used (Figure 4.9a). Using three fluorochromes, the display of seven different colors and with five fluorochromes is possible and the hybridization of the 22 human chromosomes in the same metaphase is possible. As described before, multiple probes can be labeled directly with fluorophores or using available haptens.

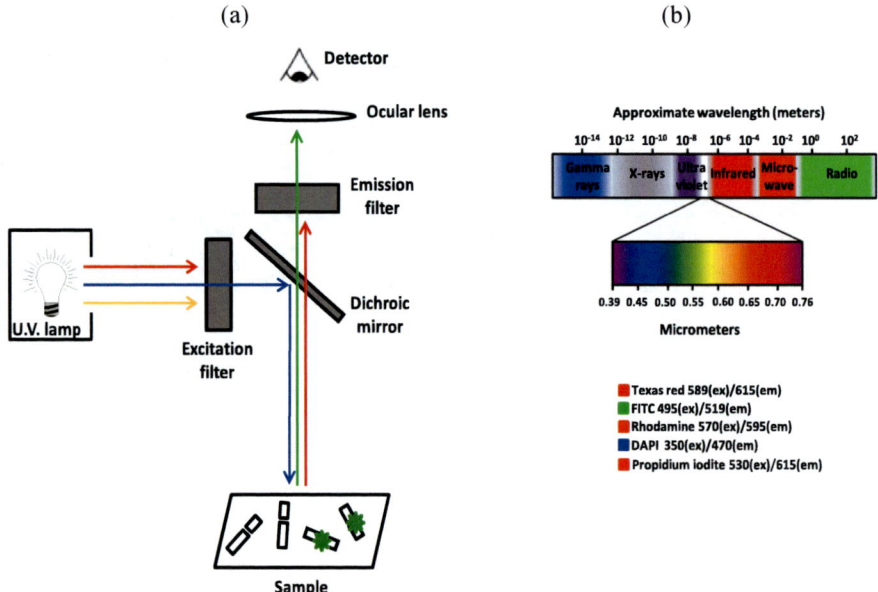

Figure 4.10. Microscope apparatus for FISH analysis. (a) internal epifluorescence system. (b) distinct wavelengths emitted by the fluorescence microscope lamp; excitation (ex) and emission (em) wavelengths for commonly used fluorochromes.

For DNA sequences that appear overlapped in FISH experiments using metaphasic chromosomes, it is useful to use distended DNA fibers. This variation in the FISH technique is named DNA combing or fiber-FISH and consists in the extension of DNA fibers before the hybridization step. It is a very useful method in assessing the length of DNA probes, and in the analysis of the organization of sequences relative to one another and interspersed sequences. This method has been extensively used in the mapping of satellite DNAs and BAC probes providing maps with higher resolution than the generated using metaphasic chromosomes, or initial meiotic cells, such as pachytene (Schwarzacher and Heslop-Harrison 2000, Speicher and Carter 2005, Jiang and Gill 2006). Sequences separated by few kilobases can be resolved using the fiber-FISH, while the powerful resolution of FISH in metaphasic chromosomes is around 5-10 Mb for barley chromosomes (Pedersen and Linde-Laursen 1995). In less condensed chromosomes observed in prophasic-prometaphasic chromosomes, the powerful resolution is increased to about 2 Mb (Cheng et al. 2002), and in pachytene chromosomes, the signal resolution can be increased approximately to 500 Kb. For smaller genomic distances, the signals appear overlapped, i.e. two probes, one green and another red will appears as a yellow signal (Schwarzacher and Heslop-Harrison 2000).

4.4. MICROSCOPE ANALYSIS FOR FISH ASSAYS

The analysis of FISH results requires the use of an epifluorescence microscope with a coupled photographic apparatus (Figure 4.10). Nowadays, the most part of images (almost all) are captured using digital cameras connected directly to a computer that facilitates the image acquisition, analysis, interpretation and increases the quality of the results for publication. The epifluorescence microscope apparatus consists basically of illumination, excitation and emission filters (that determines the wavelength to be observed), a dichroic beam-splitter (that shows the capacity of reflecting and transmission of light), and additional common parts also present in light microscopes.

Briefly, the lamp emits a light with a range of wavelengths that pass through the excitation filter (Figure 4.10). This filter is responsible for blocking all light wavelengths except the wavelengths that excite the fluorochrome that will be observed.

Then the wavelengths of interest pass to the dichroic beam-splitter reflecting some shorter light wavelengths that excite the fluorochrome linked to the probe or chromosome that will be observed. Moreover, this dichroic beam-splitter permits the transmission of longer wavelengths, which are emitted by the fluorophores in the chromosome preparations. The wavelengths that pass in the dichroic beam-splitter are filtered by the emission filter, which blocks the transmission of most wavelengths, allowing the passage of the wavelengths emitted by the fluorochrome and removing wavelengths from autofluorescence and reflected light. Finally, the light passs through the ocular lens or goes to the capture system (Figura 4.10a).

The process of signal emission (fluorescence emission) by the fluorochrome bounded to the probe or to the chromosomes consists basically of three steps. When an appropriate wavelength is directed to the chromosome slide, a specific fluorochrome absorbs one photon (step one) and changes its state, becoming excited (step two). After this process, the emission of one photon (step three) results in the fluororescent light emission that will be visualized. For a precision process, the presence of specific filters in the microscope for the fluorochromes of interest that are responsible to filter the wavelengths emitted by the microscope lamp and fluorochromes is necessary. The fluorochromes possess distinct wavelengths for excitation and emission, which may be similar between the distinct fluorochromes (Figure 4.10b). The use of more specific filters (that permits only the passage of specific wavelengths) results in FISH analysis with more quality and less background in the results. For more information about fluorochromes and technical approaches concerning the epifluorescence microscope and capture systems for FISH analysis, it is useful to check the available catalogues of some distinct manufactures.

All approaches presented along this section have been extensively used in cytogenetics and have provided a good resolution for studies related to chromosomal evolution and genome organization. Some examples of the application of these methodologies using repetitive DNAs, single copy sequences and chromosome paints, for analysis of normal chromosomal complement, including autosomes and sex chromosomes, and supernumerary elements will be presented in the next sections of this book, showing the importance and power of the FISH technique for studies of karyotypic tracking in animal chromosomes.

Chapter 5

REPETITIVE DNAS: MORE THAN JUST "JUNK" DNA

5.1. BASIC FEATURES OF REPETITIVE DNAS

The presence of a great amount of repetitive sequences is a common feature of eukaryotic genomes. These sequences are characterized by high variability and constitute families of repeated DNAs that represent a substantial component of eukaryote genomes, in some cases more than 80% of the DNA content of the cell (Charlesworth et al. 1994, Ridley 1996, Gregory 2005b, Plohl et al. 2008) (Figure 5.1). The large variation in genome size reported for eukaryotes is attributed, mostly, to the accumulation of repeated sequences (Petrov 2001, Kidwell 2002). The repeated DNA may vary between species with no obvious relation to the complexity of the organism, number of genes or ploidy level (Gregory 2005b). Briefly, the accumulation process involves repeated DNA amplification and may occur primarily via gene conversion, unequal crossing-over, transposition and RNA-mediated transfer (Charlesworth et al. 1994, Hancock 1999, Kidwell 2005) (Figure 5.2).

Repetitive DNA consists of sequences, identical or similar, which are in tandem (side by side) or dispersed throughout the genome (Figure 5.3). The satellite DNAs (satDNAs) represent a typical example of tandem array repeated sequences whereas the transposable elements (TE) represent scattered sequences, although sometimes TEs can be organized in blocks in the chromosomes (Long and Dawid 1980, Charlesworth et al. 1994, Sumner 2003). Although today it is known that transposable elements and satDNAs can be transcribed, for a long time the general consensus established these sequence as non-encoding (Figure 5.1). The "non-encoding" sequences are represented by satDNAs, minisatellite, microsatellite and transposable elements and the encoding ones are composed by the multigene families composed of hundreds to thousands of copies, such as the histone and ribosomal RNA (rRNA) genes, among others (Kedes 1979, Flavel 1986, Prokopowich et al. 2003, Nagoda et al. 2005). For too long, the sequences of repeated DNAs were considered "junk DNA" due to the lack of knowledge of their functions (Nowak 1994, Biémont and Vieira 2006). However, several studies have suggested, and often demonstrated, that this type of DNA is involved in structural and functional organization of the genome, acting on gene regulation, replication, DNA repair, chromosomal rearrangements, differentiation and chromosome segregation and may also be involved with diseases.

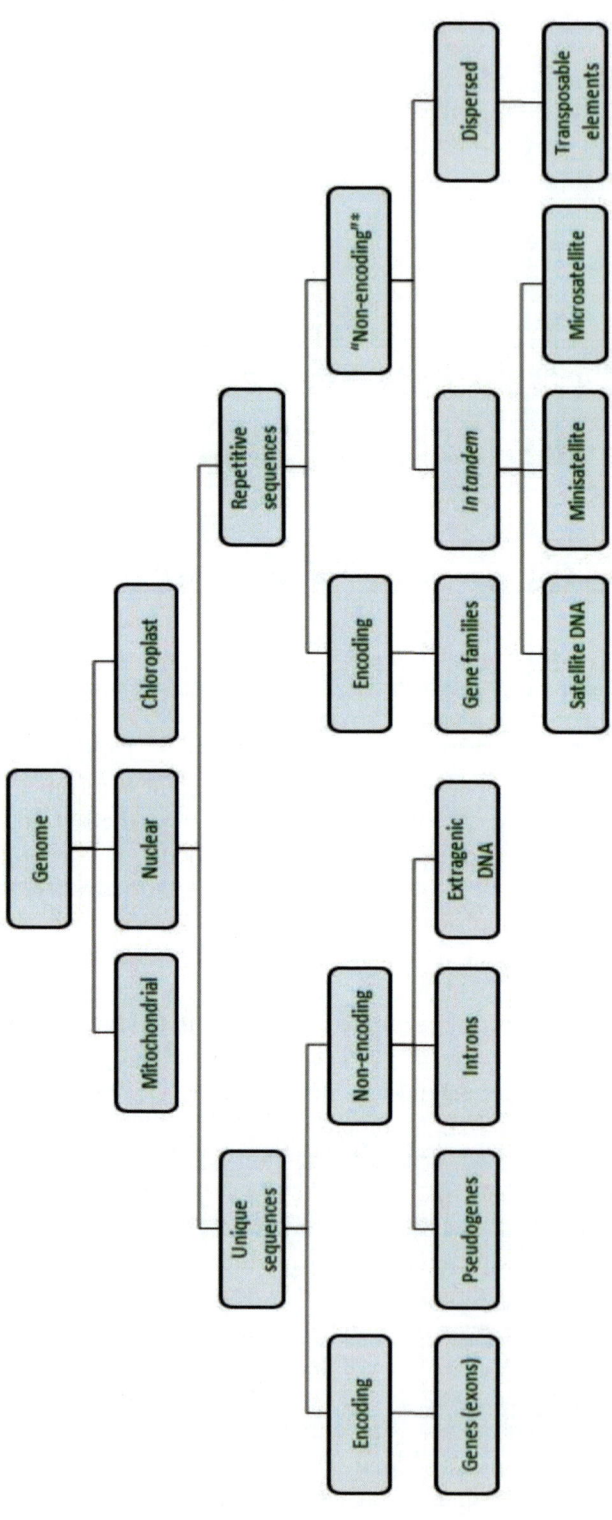

Figure 5.1. Scheme of genome organization of different classes of DNA sequences in the eukaryotes. The "non-encoding" nature of repetitive sequences (*) is explained in the text.

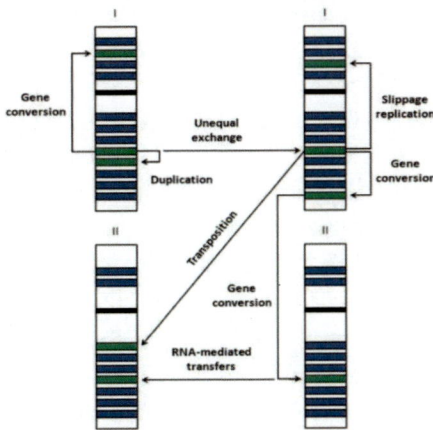

Figure 5.2. Evolutionary dynamics of repetitive sequences in the genome. I and II represent different chromosome pairs with their homologous representatives.

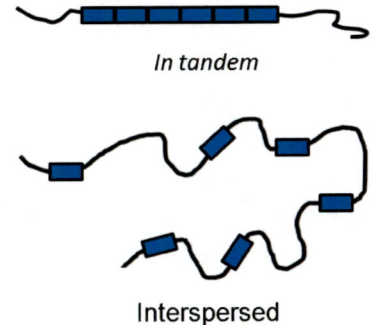

Figure 5.3. Scheme of repetitive DNA organization in tandem or dispersed through the genome.

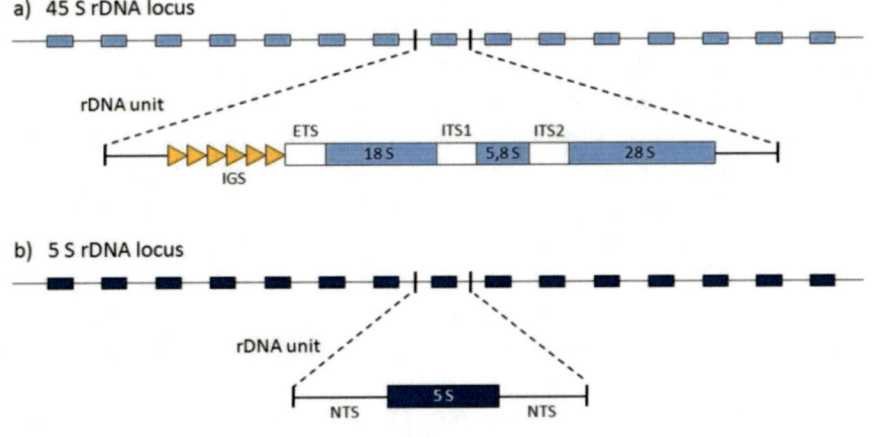

Figure 5.4. Organization of ribosomal RNA (rRNA) genes in eukaryotes. The genes are organized into tandemly repeated units as diagrammed at the top. A typical unit is shown in expanded detail. IGS, intergenic spacer; ETS, external transcribed spacer; ITS, internal transcribed spacer; NTS, nontranscribed spacer.

Recently, the large-scale sequencing of genomes has produced an unprecedented wealth of information about the origin, diversity, and genomic impact of the repetitive sequences. Repetitive DNAs represent great chromosomal markers that are very useful in studies of chromosome structure and function, evolution, identification of chromosomal rearrangements, supernumerary chromosomes and sex chromosomes. Fluorescence *in situ* hybridization, using repetitive DNAs as probes, is a powerful technique that can correlate molecular information of a DNA sequence with its physical location along chromosomes and genomes, and has been applied to chromosomes in various eukaryotes, including several animal groups (Fridolfsson et al. 1998, Nomoto et al. 2001, Vitturi et al. 2002, Volpi and Bridger 2008, Mazzuchelli and Martins 2009, Veltsos et al. 2009, Cabral-de-Mello et al. 2010a) (Figure 3a-d, j-l).

The application of fluorescence *in situ* hybridization for repeated DNAs in animals frequently use rDNAs, transposable elements or satDNA as probes, but there are also studies using histone genes and other sequences. The rDNAs and some satDNAs have been mapped in all vertebrate groups and in several invertebrates, such as annelids, mollusks, arthropods and echinoderms, whereas the histone genes have only been mapped, for example, in insects (Hankeln et al. 1993, Ranz et al. 2003, Eirín-López et al. 2004, Cabrero et al. 2009, Cabral-de-Mello et al. 2010 a, b), mollusks (Zhang et al. 2007), fishes (Pendás et al. 1994), amphibians (Turner et al. 1988), and mammals (Graves et al. 1985, Tripputi et al. 1986).

5.2. MULTIGENE FAMILIES FOR RIBOSOMAL RNA AND HISTONES

Most multi-gene families physically mapped on chromosomes by fluorescence *in situ* hybridization are the rRNAs and histone genes. Ribosomal DNA (rDNA) transcribes the rRNA component of ribosomes, which are essential for the protein synthesis process. Eukaryote genomes contain multiple copies of rRNA genes, presumably because exceptionally high quantities of RNA transcripts are necessary. These genes are organized in tandem arrays containing transcriptional units for 18S, 5,8S and 28S ribosomal RNAs transcribed from a single promoter by RNA polymerase I. The repeat units containing the transcribing sequences for the 18S, 5,8S and 28S ribosomal RNAs are separated from each other by variable intergenic spacers (IGS) and an external transcribed spacer (ETS) (Eickbush and Eickbush 2007) (Figure 5.4a). Ribosomal DNA repeat units are evolutionary dynamic regions of the genome and seem to be able to spread through the genome creating new rDNA loci. The rDNA tandem arrays evolution is thought to occur through concerted evolution mechanisms and the sequence's high-similarity is maintained by the homogenization of rDNA repeat arrays (by homologous recombination between tandem repeats). In eukaryotes, 5S rRNA is also transcribed from tandemly repeated sequences, but are often of higher copy number than the others rRNA genes. The 5S rDNA repeats consist of a highly conserved transcribed sequence of 120 bp, approximately, which is separated from each other by a variable non-transcribed spacer (NTS), transcribed by RNA polymerase III, the polymerase is also responsible for the synthesis of tRNAs and other small, non-translated RNAs (Bell et al. 1977, Martins and Wasko 2004, Haeusler and Engelke 2006, Richard et al. 2008) (Figure 5.4b).

Most cytogenetic studies using rDNA as probes in invertebrates have been conducted in insects, primarily using the 45S rDNA, and some variation for the chromosomal distribution of this marker has been described (Sánchez-Gea et al. 2000, Cabrero and Camacho 2008, Nguyen et al. 2010). Although some studies concerning the rDNA sites have been extensively conducted by means of silver nitrate staining, detecting only active nucleolar organizer regions (NOR), that can not correspond to the real genome organization of 45S rDNA clusters (for example, see Cabrero and Camacho 2008, Cabral-de-Mello et al. 2010a). Concerning the 5S rDNA, there are few reports in the literature for chromosomal location of this gene in grasshoppers, beetles and other insects (Bizzaro et al. 2000, Cabrero et al. 2003a, Loreto et al. 2008, Cabral-de-Mello et al. 2010a, c). In grasshoppers, this sequence was mapped in *Eyprepocnemis plorans* (Cabrero et al. 2003a), *Rammathocerus brasiliensis* (Loreto et al. 2008), *Locusta migratoria* (Teruel et al. 2010) and in four Proscopiidae species (Cabral-de-Mello et al. 2010c). The first three species presented some clusters of 5S rDNA, while in the Proscopiidae species, a remarkable conservation for this sequence was observed in both number and location of sites. Moreover, this sequence was observed in the B chromosome of *E. plorans* and in *R. brasiliensis,* being an important marker to analyze B chromosome origin and evolution (Cabrero et al. 2003a, Loreto et al. 2008). The 5S rDNA was located only in one species of Coleoptera and one of Hemiptera. In *Dichotomius geminatus* (Coleoptera), this sequence was located in one chromosome distinct of the location of 18S rDNA, whereas in *Acyrthosiphon pisum* (Hemiptera) the two rDNA clusters were located in the same chromosome but in distinct chromosome regions (Bizzaro et al. 2000, Cabral-de-Mello et al. 2010a).

Among other invertebrates, the 18S-28S and 5S rDNAs were mapped primarily in Mollusca, Annelida and Crustacea, and were frequently detected in conspicuous blocks in the same chromosome or in separated chromosomes (Drouin and Moniz de Sá 1995, Colomba et al. 2002, Vitturi et al. 2002, 2004, Wang and Guo 2004). In the mollusk *Melarhaphe neritoides* (Caenogastropoda), for example, the repeated units of both rDNA classes are closely associated on the same chromosome, most probably interspersed due to the overlapping of the two hybridization signals (Colomba et al. 2002). Similar pattern was also observed in Annelida, some other mollusks and crustaceans (Drouin and Moniz de Sá 1995, Vitturi et al. 2002, 2004). However, in some other species, a distinct scenario with isolation of these two multi-gene families was also reported, such as in the beetle *Dichotomius geminatus*, and some mollusks (Cabral-de-Mello et al., 2010a, Insua et al. 2006, Huang et al. 2007).

For vertebrates, many studies for the rDNA cluster were performed in fish, and the scenario of rDNA clusters organization indicates a non-association of the two rDNA classes, with most species presenting these sequences in distinct chromosomes. On the other hand, for some species, both rDNA classes are clustered near each other or far in the same chromosome (Lucchini et al. 1993, Mandrioli et al. 2000, Sola et al. 2000, Martins and Galetti 2001, Pisano and Ghigliotti 2009). In general, 5S rDNA occurs in interstitial regions of the chromosomes (Lucchini et al. 1993, Martins and Wasko 2004, Ferreira et al. 2007, Gornung et al. 2008), and apparently this pattern may represent an ancestral condition or could confer some advantage for the genome organization of these sequences (Martins and Galetti Jr 1999, 2000). Another interesting characteristic described for 5S rDNA in fishes is the presence of two distinct sequence classes organized in different chromosomal regions or even in different chromosomes (Sajdak et al. 1998, Martins et al. 2000, Martins and Galetti 2001).

The variations observed for rDNA sites indicate a complex microevolutionary pattern that rules their organization in the genome. Ribosomal DNAs seem to be able to spread through the genome, thus creating new rDNA loci (revised by Raskina et al. 2008), variant rDNA copies (Martins et al. 2006) and even association to other multi-gene families (Eirín-López et al. 2004, Cabral-de-Mello 2010a, c, Drouin and Moniz de Sá 1995). Several spreading mechanisms have been suggested, including transposition, insertion of extrachromosomal rDNA amplified during oogenesis, the presence of repetitive elements facilitating nonhomologous chromosome exchange, and the amplification of minor rDNA loci (Phillips et al. 1988, Maggini et al. 1991, Dubcovsky and Dvořak 1995).

The second multi-gene family most explored by fluorescence *in situ* hybridization codifies the histones and is known to contain moderately repeated genes (Kedes 1979). The histone genes have an extraordinary organization in tandem arrays, are interspersed from each other with noncoding spacer sequences, and codifies for five histone proteins (H1, H2A, H2B, H3, and H4) (Kedes and Gross 1969, Kedes 1979, Barzotti et al. 2000, Nagoda et al. 2005) (Figure 5.5). This pattern has been observed in several organisms such as *Chaetopterus variopedatus* (del Gaudio et al. 1998), *Drosophila melanogaster* (Lifton et al. 1978), and *Salmo gairdnerii* (Connor et al. 1984). It was discovered in some genomes a few atypical unit lacking every gene (Schienman et al. 1998, Nagoda et al. 2005). In some organisms, such as *Gallus domesticus* (Engel and Dodgson, 1981), mouse (Sittman et al. 1981) and human (Albig et al. 1991), it has been observed that these genes may also be encoded as single genes or groups of genes. The organization and arrangement of histone genes has undergone expressive evolutionary changes, but it is not yet clear how these different units originated and spread in the genome (Nagoda et al. 2005). The structural differences in the histone repeat units among diverse organisms suggests they originated by evolutionary amplification of a unique ancestral histone gene cluster. Alternatively, a single ancestral cluster may have undergone various rearrangements and sequence changes to produce the diversity of fine structures in the histone gene clusters found in different organisms (Maxson et al. 1983, Schienman et al. 1998).

Histones are a class of basic proteins that associate with each other and with nuclear DNA to form the nucleosome, the fundamental unit of chromatin structure (see Chapter I). The structure of histones, particularly the H3 and H4 histones, is generally highly conserved between diverse animal phyla and even between the animal and plant kingdoms, although remarkable variations have been reported in the DNA sequence that codes for these proteins (Ruberti et al. 1982, Maxson et al. 1983, Miller et al. 1993, del Gaudio et al. 1998, Tsunemoto and Matsuo 2001, Albig et al. 2003). These proteins are sites of some pos-translational modifications, which include acetylation, methylation, phosphorylation, ADP-ribosylation, and ubiquitination, playing a fundamental role in the precise regulation of the chromatin structure and function (Goll and Bestor 2002, Kouzarides 2007).

Figure 5.5. Histone gene arrangements in the quintet repeating unit.

The use of histone genes in chromosomal mapping have been mostly conducted in invertebrates being the analysis concentrated in midges, fruit flies, mollusks, grasshoppers and only one beetle (Hankeln et al. 1993, Schienman et al. 1998, Ranz et al. 2003, Eirín-López et al. 2004, Zhang et al. 2007, Cabrero et al. 2009, Cabral-de-Mello et al. 2010a, c). In general, these sequences have presented more conservation related to chromosomal location and number of clusters compared to the rDNAs (Cabrero et al. 2009, Cabral-de-Mello et al. 2010a), although low variability in cluster number has been reported (Ranz et al. 2003, Zhang et al. 2007). The recent study in grasshopper performed by Cabrero et al. (2009) revealed a strong association of H3 and H4 histones in grasshoppers and intense conservation of chromosomal location of these sequences. This chromosomal conservation was also described for other insects (Cabral-de-Mello et al. 2010a, c). Recently, Teruel et al. (2010) used nucleotide sequences of histones for the establishment of origin of B chromosome in the migratory locust *Locusta migratoria*. Some molecular analyses for histone gene sequences have revealed that these sequences form a block of tandem arrays in some groups. This pattern of histone organization has been detected for some invertebrates as cited above and in three vertebrate fish species, *Salmo salar* (salmon), *S. trutta* (brown trout) and *Oncorhynchus mykis* (rainbow trout) (Pendás et al. 1994). Moreover, histone sequences and 5S rRNA genes can be linked in the same cluster in animal genomes. This finding was reported in one mussel and two crustaceans (Drouin and Moniz de Sá 1995, Barzotti et al. 2000, Eirín-López et al. 2004). Moreover, the FISH using probes for 5S and H3 histone genes in beetles (Cabral-de-Mello et al. 2010a) and four ancient grasshoppers (Cabral-de-Mello et al. 2010c) revealed a unique block for these two sequences, indicating a possible interspersed association of both genes.

5.3. IN TANDEM REPEATED DNAS

5.3.1. General Features of in Tandem Repeated DNAs

The classification of repeated sequences organized in tandem is based on the repeated unit and cluster repetition sizes. The first group is represented by the satellite DNA (satDNA), composed of highly repeated sequences, grouped into one to several locations, along one or more chromosomes and interspersed with single copy sequences. The repeated units range from 100 to 1.000 nucleotides, varying in structure, location and quantity within the genomes (Ridley 1996, Ugarković and Plohl 2002). Longer repeat size (even longer than 4 kb) can also be present in the genome and are named megasatellite DNAs (Gondo et al. 1998). SatDNAs represent the main component of the heterochromatin and are located preferentially in pericentromeric and telomeric regions (Yunis and Yasmineh 1971, John and Miklos 1979, Juan et al. 1993, Shapiro and Sternberg 2005). SatDNAs generally show high variability in monomer size, nucleotide sequence, copy number and chromosomal organization (Charlesworth et al. 1994). Some of the satDNA families evolve quickly and allow their use as molecular markers in comparative studies (Watabe et al. 1997, Picariello et al. 2002, Kuhn and Sene 2005, Franco et al. 2010).

The second group consists of moderately repeated DNA, with repeat units about 10 to 100 bp, called minisatellites or sequences with variable number of repeats (VNTR - Variable

Number of Tandem Repeats) (Jeffreys et al. 1985). Each cluster of repeats represents a minisatellite locus, whose alleles are distinguished by variations in their overall size and can be used as genetic markers in a technique called DNA fingerprinting (Mueller and Wolfenbarger 1999). Minisatellites have been proposed as intermediates in chromosome pairing initiation in some eukaryote genomes (Ashley 1994, Sybenga 1999), which might be related to their proposed genetic recombination properties (Boan et al. 1998, Wahls and Moore 1998). The distribution is, however, very different, with a high bias toward chromosome ends, and a closer look at the syntenic relationship suggests that, in a common ancestor, the interstitial minisatellite clusters were telomeric (Amarger et al. 1998, Vergnaud and Denoeud, 2000). One of the first minisatellites was described in an intron of the human myoglobin gene and is comprised of 33 bp tandem repeat units with some sequence similarities with other minisatellites discovered previously. It is flanked by a 9 bp direct repeat, a characteristic signature of transposable elements, suggesting that this minisatellite was able to transpose in some way (Jeffreys et al. 1985). The minisatellites were further identified in a wide variety of organisms, such *Arabidopsis thaliana, Caenorhabditis elegans, Drosophila melanogaster, Homo sapiens* and *Saccharomyces cerevisiae*, among others, demonstrating that such DNA class is common in the eukaryote genomes (Richard et al. 2008).

Another group of tandem repeated DNAs is composed of short repeats of 1 to 6 bp named microsatellites (also known as simple sequence repeats - SSRs) (Trivedi 2004). They are widely distributed in the genome, highly polymorphic and therefore used as markers in studies of population genetics, conservation, epidemiology, testing and kinship mapping (Bowcock et al. 1994, Balloux et al. 1998, Röder et al. 1998, Schlötterer 2000). Among the functions assigned to microsatellites are their participation in chromatin organization, DNA replication, recombination and regulation of gene activity (Li et al. 2002a). The first microsatellite was characterized by Weller and colleagues as a polymorphic $(GGAT)_{165}$ repeat in the human myoglobin gene (Weller et al. 1984, Richard et al. 2008). Another example is the telomeric repeats, which are composed of the hexameric unit TTAGGG in all mammals and other animals, and are essential for chromosome stability and regulation of replicative lifespan of somatic cells (Blackburn 2005, Fouché et al. 2006). Some variability for telomeric sequence has also been reported for animals and the abundance of microsatellite into telomeres was observed from protozoa to mammals (Katzen et al. 1981, Blackburn and Challoner 1984, Ponzi et al. 1985, Baroin et al. 1987, Sugawara and Szostak 1986, Forney et al. 1987, Richards and Ausubel 1988, Allshire et al. 1989).

Microsatellites were observed in a range of organisms, such as *Escherichia coli, Haemophilus influenzae, Helicobacter pylori, Arabidopsis thaliana, Drosophila melanogaster, Homo sapiens, Mus musculus, Saccharomyces cerevisiae, Schizosaccharomyces pombe* and *Tetraodon nigroviridis* (Field and Wills 1998, Richard et al. 2008), being common and widespread in the prokaryote and eukaryote genomes. Microsatellites can also be associated to non-LTR retrotransposons as observed in the lepidopteran genomes. Slippage during reverse transcription of retrotransposon and chromosomal integration can increase the number of repeating units within the SSR, and these events can account for a substantial fraction of unilateral repeat families recovered in independent efforts (Tay et al. 2010).

5.3.2. Cytogenetic Mapping of Tandem Repeated DNAs

Despite the large number of studies of in tandem repeated DNA, the cytogenetic mapping of microsatellites and minisatellites is scarce, and the satDNA is the best characterized group of sequences. There are several studies in some animal groups related to chromosomal location, molecular organization and possible functions of heterochromatin (Sumner 2003, Grewal and Jia 2007). However, despite the heterochromatin having been extensively studied, the mapping of satDNAs is restricted to a few groups (Phillips and Reed 1996, Odierna et al. 2004, Wang et al. 2001, Palomeque and Lorite 2008, Giovannotti et al. 2009). Usually satDNA families are species-specific (Arnheim 1983). However, there are exceptions in which a group of species share the same satDNA family, as observed for the centromeric alpha satDNA, which is preserved in the primate order most probably because of its centromeric function (Schueler et al. 2001). Alpha satellite-like sequences were also detected in other organisms, as chicken, primates and zebrafish (Li and Kirby 2003, Alkan et al. 2007).

Despite their sequence heterogeneity, the study of insect satDNAs indicates the evolutionary conservation of certain features (such as total length, monomer length, motifs, particular regions and/or secondary and tertiary structures), and within each taxonomic group, the satDNAs have been studied in only a few species (reviewed in Palomeque and Lorite 2008). SatDNA has been characterized in eight species of the *Formica* genus (Hymenoptera, Formicidae). This satDNA is organized as tandem repeats of 129 bp monomers (with the presence of internal inverted repeats) and is located in the pericentromeric regions of all chromosomes (Lorite et al. 2004). Although the evolution of the satDNA in ants could be similar to that in other organisms, there are some particularities as a result of a haplodiploid system. Ant male haploids do not undergo meiotic recombination and redistribution of chromosomes to the next generation. In Hymenoptera, the processes of molecular evolution of the satDNA would be altered by the special constraints imposed by the haplodiploid system (Bigot et al. 1990). It was suggested that the satellite dynamic appears to be the outcome of both general molecular processes and specific organism traits (Luchetti et al. 2003).

Other examples of the use of satDNA in chromosome and genome structure studies in insects were performed in beetles, for example, belonging to Tenebrionidae family and in grasshoppers for analysis of B chromosome structure (Mravinac et al. 2004). In tenebrionids that present species with high amount of AT rich centromeric heterochromatin, the mapping of satDNA was applied in chromosomes of representatives species from the genus *Alphitobius, Misolampus, Palorus, Tenebrio* and *Tribolium*, and the results were similar to those observed in other groups with distinct satDNAs located in heterochromatic regions (Mravinac et al. 2004, Mravinac et al. 2005). For the genus *Tribolium*, the satDNA families have conserved and variable segments and common characteristics, as short inverted repeats, non-random distribution of A or T$\geq$3 tracts and a CENP-B box –like motif, and may present functional elements for specific protein recognition and condensation of heterochromatin (Mravinac et al. 2005).

For grasshoppers, one satDNA with 180 bp was mapped in distinct population of the B chromosome harboring species *Eyprepocnemis plorans*. This sequence was present in the centromeric region of some autosomes and X chromosome with variations between the populations. Moreover, this sequence was mapped in B chromosomes of some populations and showed variability related to the presence/absence of this sequence that interestingly, was correlated with the distinct populations (Cabrero et al. 2003a, Abdelaziz et al. 2007). On the other hand, this sequence was not detected in chromosomes of other species of the same genus and other closely relative species by FISH analysis, indicating that it is probably specific for *E. plorans* (Cabrero et al. 2003b, López-León et al. 2008).

The physical mapping of satDNAs in mollusks provided an effective tool for oyster chromosome identification which is so difficult due their small size and similar arm ratio (Guo et al. 1996). Clabby et al. (1996) identified a highly repetitive satDNA, named *Cg*170, in the Pacific oyster. *Cg*170 was the first centromeric satellite sequence identified in bivalves and in mollusks in general. Other satellite sequences have been reported in a few mollusks, which are apparently unrelated to *Cg*170 satellite, and provide important insights on the organization and evolution of mollusk genomes (Plohl and Cornudella 1997, Ruiz-Lara et al. 1992, Mary-Elizabeth et al. 1998, Wang et al. 2001, Petrović et al. 2009). In *Donax trunculus*, a highly abundant satDNA named DTF2 was mapped on chromosomes and overlapped only partially with the GC-rich heterochromatin located in interstitial and subtelomeric regions (Petrović et al. 2009).

For vertebrates, satDNAs were mapped mainly in fish and mammal chromosomes, although there are examples concerning amphibians, birds and reptiles. The chromosomal distribution of satDNAs among fish species showed that these sequences are mainly located in the centromeric region of chromosomes. A *Hind*III satDNA family isolated from the sturgeon *Acipenser naccarii* genome was preserved in the pericentromeric regions of the chromosomes of six species of the genus *Acipenser* and one of the genus *Huso* (Lanfredi et al. 2001). Centromeric satDNA families were also isolated from the genomes of the gobiid *Gobius cobitis* (Canapa et al. 2002) and the Nile tilapia *Oreochromis niloticus* (Ferreira and Martins 2008). Particularly in the Nile tilapia, the satellite family was present in the centromeric region of all chromosomes of the complement (Ferreira and Martins 2008), suggesting that this satellite sequence is derived from an ancestor of the group, and has been maintained in the centromeres of all chromosomes due to its functionality. An interesting example that illustrates the intense dynamics of satDNAs was observed in the 5S*Hind*III satDNA of the fish *Hoplias malabaricus*, which share similarities to 5S rDNA true repeats (Martins et al. 2006). The high evolutionary dynamic of 5S*Hind*III makes this sequence useful in the identification of chromosome rearrangements, sex chromosomes and populational variations in the species (Ferreira et al. 2007, Cioffi et al. 2009).

Particular attention has been directed to the identification of satDNAs related to sex and supernumerary chromosomes in fish (see Chapter VIII). SatDNAs have been isolated and mapped in sex chromosomes of several species, such as *Leporinus elongatus* (Nakayama et al. 1994), *Chiondraco hamatus* (Capriglione et al. 1994a), *Poecilia reticulata* (Nanda et al. 1990), *Oncorhynchus tschawytscha* (Devlin et al. 1991, Stein et al. 2001), among others; offering contribution to the knowledge of sex chromosomes evolution in fish. A satDNA, named As51, correlated to a supernumerary chromosome was firstly isolated in *Astyanax scabripinnis*, this element had repeats of 51 bp and was located in the non-centromeric heterochromatins, in the NOR regions, and in the supernumerary chromosome. The symmetric distribution of As51 in both arms of the B chromosome and its meiotic behavior suggest that this chromosome is an isochromosome (Mestriner et al. 2000). SatDNAs were also isolated from *Prochilodus lineatus* which presents 0-5 small supernumerary chromosomes. Two satDNA families, with monomeric units of 441 and 900 bp, were isolated from the genome of this species and were located in the pericentromeric region of several chromosomes of the A complement. The 900 bp satellite was also located in several supernumerary chromosomes, demonstrating that these supernumerary elements have originated from A chromosomes that harbor the 900 bp satDNA family (Jesus et al. 2003).

In amphibians, for example, studies of the *Pst*I satDNA family contributed to understanding the origin of tetraploidy in *Bufo viridis* (Anura), suggesting that this was an ancient event (Odierna et al. 2004). In green toads, a group of frogs, the diversification appears to occur without any chromatin change, including the localization and composition of heterochromatin and of the *Pst*I satDNA family (Birstein and Mazin 1982, Matsui et al. 1995, Odierna et al. 2004). In *Discoglossus pictus* (Anura), it was observed that there is an unusual chromosomal distribution of the *Dp-sat1* satellite DNA. This satDNA is characteristic of the genus *Discoglossus* and represents a major repetitive DNA that accounts for about 6% of the *D. pictus* genome. Two interesting aspects are the non-random distribution of the *Dp-sat1* DNA and its association in the interphasic nucleus. Most *Dp-sat1* associated heterochromatin appears to be located in only one pole of the nucleus, showing evidence of ordered organization and location of the chromosomes in the nucleus during interphase (Amor et al. 2009).

There are few studies in reptiles with satDNAs, and usually they are related to molecular analysis without cytogenetic mapping (Caputo et al. 1994, Grechko et al. 2005, Giovannotti et al. 2009). A large number of satDNAs sequences are of recent origin and are only observed in related species (Capriglione et al. 1994b, Rudykh et al. 1999, Ciobanu et al. 2003, 2004, Grechko et al. 2005), whereas the most common ones are also highly conserved in unrelated species, maybe due to any important function (Olmo et al. 2002). The first satDNA isolated and characterized in Scincidae (Reptilia) was the *Ava*II satDNA isolated from the genome of *Eumeces schneideri*, one of the most diverse and species-rich families of squamate reptiles (Giovannotti et al. 2009). The mapping of this sequence revealed that this satellite is widespread in the genome of this lizard, occurring on most of the 16 chromosomes of the species with variation in signal intensity. Moreover, the fiber-FISH using *Ava*II satDNA and telomeric sequence showed an overlapping of these sequences. Comparative nucleotide sequence analyses of *Ava*II satDNA with sequences of satDNAs deposited in GenBank revealed no similarity, indicating that *Ava*II satDNA constitutes a new satellite family (King and Cummings 1997). In *S. scincus*, considered the most closely species to *E. scheneideri*, it was also observed the presence of the *Ava*II satDNA. The occurrence of this satellite in phyletic lineages which split about 14 million years ago (Carranza et al. 2008), and the presence of a unique satellite family in *E. schneideri* indicates that these taxa are considered as chromosomal conservative lineages (Slamovits and Rossi 2002). Through the karyotype stability observed in *E. schneidei* and the closely related species *E. algeriensis*, *S. scincus*, and *S. hemprichii,* this hypothesis seems to be confirmed (Caputo et al. 1994).

One characteristic of the centromeric, highly repetitive DNA sequence associated with centromeric heterochromatin is the genomic compartmentalization between macro- and microchromosomes. The results obtained in two primitive bird orders (Galliformes and Struthioniformes) suggest that the homogenization of the centromeric repetitive sequences has not occurred between macro- and microchromosomes owing to their structural differences (Matzke et al. 1990, Matzke et al. 1992, Tanaka et al. 2000, Yamada et al. 2002a, b). In turn, the *Crocodylus siamensis* (Crocodylidae) is a good model to examine the correlation between karyotypes and chromosomal size-dependent distribution of centromeric repetitive sequences. The kayotype of *C. siamensis* has 2n = 30 comprised predominantly by bi-armed chromosomes with no dot-shaped microchromosomes. The centromeric heterochromatin of *C. siamensis* is composed of two major components of repetitive sequence families, and they exhibited remarkable differences in structure, genomic organization, chromosomal distribution and conservation of nucleotide sequences. Therefore, the two centromere-specific sequence families were not co-localized in the same regions of centromeric heterochromatin, revealing the compartmentalization of centromeric repetitive sequences in the genome (Kawagoshi et al. 2008).

The amount of pericentromeric satDNA is highly variable in mammals. In several species of the genus *Acomys* (Rodentia), for example, a family of centromeric satDNA represents about 0.88% of the haploid genome. In mice, the less abundant satDNA described until now represents about 0.5-1%, and the higher abundant represent approximately 6% of the genome (Kunze et al. 1999). Mammalian centromeres are composed of multiple families of satDNA, and the centromeric satellites are well-characterized in humans and other primates and are usually characterized by the presence of tandem repeats of monomers from 169 to 170 bp, which are organized into chromosome -specific high-order repeat units (Willard and Waye 1987, Choo et al. 1991, Wang et al. 2001).

The majority of satDNA families have an important role in mammal genome evolution by promoting chromosomal rearrangements (Wichman et al. 1991, Qumsiyeh 1994, Slamovits et al. 2001, Louzada et al. 2008, Ropiquet et al. 2008). The involvement of constitutive heterochromatin in events of chromosomal evolution is discussed in several studies, suggesting that these regions act as hotspots enabling structural chromosome rearrangements (Chaves et al. 2004, Louzada et al. 2008, Ventura et al. 2009). The location of evolutionary breakpoint regions is coincident with the location of regions rich in repetitive sequences (Li et al. 2000, Locke et al. 2003, Ruiz-Herrera et al. 2006). The Indian muntjac (*Muntiacus muntjak vaginalis*) has a karyotype of 2n = 6 in the female and 2n = 7 in the male (Wurster and Benirschke 1970), and is a classic example of occurrence of extensive chromosome rearrangement. The karyotype evolution of Indian muntjac via extensive tandem fusions and several centric fusions, besides the occurrence of other rearrangements, was well-documented by molecular cytogenetic studies (Yang et al. 1995, Chi et al. 2005, and others). It has been characterized in the Indian muntjac three cervidae-specific centromeric satDNA sequences (Bogenberger et al. 1987, Lin et al. 1991, Qureshi and Blake 1995, Vafa et al. 1999, Li et al. 2002b, Liu et al. 2008). A segment of centromeric DNA identified revealed a complex organization of sequences, and only two regions have sequences homologous to previously identified cervid centromeric satDNA families, also showing similarities to interspersed repetitive sequences and sequences scattered over this segment of centromeric DNA. It might be that Indian muntjac centromeric DNA segments represent characteristics of the centromere of Cervidae or is the result of chromosome restructuring during the process of karyotypic differentiation (Cheng et al. 2009).

Another example of the involvement of satDNA in chromosomal diversification was described in *Equus* (Mammalia) involved with neo-centromere formation. In this genus, the centromere repositioning (the shift along the chromosome of the centromeric function without DNA sequence rearrangement) has occurred frequently during the evolution. It has been proposed that the initial event of evolutionary repositioning may be the loss of function of the original centromere followed by the gain of epigenetic signals in a non-centromeric position. Marshall et al. (2008) proposed that satellite sequences are incorporated at repositioned centromere sites, because they probably confer an adaptive advantage, possibly by increasing the accuracy of chromosome segregation. Alternatively, the accumulation of satellite sequences may simply be a neutral process driven by the presence of heterochromatin in the centromeric DNA. In the horse (*Equus caballus*), the satDNA colocalize with centromeric protein CENP-A, with the exception of chromosome 11 in which CENP-A, but not satDNA, is present at the centromere; in the donkey (*Equus asinus*), CENP-A is present at all centromeres (primary constrictions) but satellite signals are often absent at these sites while present at several non- centromeric ends (Piras et al. 2010). The centromeric function and the position of satDNA turned out to be often uncoupled. Piras et al. (2010) suggests the satellite-less centromeres arose from two different evolutionary events: fusions between ancestral acrocentric chromosomes and centromere repositioning. Satellite repeats may colonize new centromeres at a later stage giving rise to "mature" centromeres by a mechanism of neocentromere formation during the evolution.

Although minisatellites have been described for many groups, the chromosomal mapping of these sequences was only described for a few species (Rabenold et al. 1990, Estoup et al. 1993, Ellegren et al. 1994, Naciri et al. 1995, Fischer et al. 2000, 2004, Santos et al., 2010). In fish, only few analyses of minisatellites mapping have been conducted. Three were mapped minisatellites with core sequences of 42, 28 and 34 nucleotides in the chromosomes of *Salmo salar* and this was the first cytogenetic mapping in the Atlantic salmon (Pérez et al. 1999). Concerning the distribution of minisatellite sequences in the genome of the pufferfish (*Tetraodon nigroviridis*), two minisatellites were identified, ms100 and ms104 (minisatellite units containing repeats of 104 and 100 bp in length, respectively) (Crollius et al. 2000, Fischer et al 2004), that were visualized predominantly in heterochromatic regions, suggesting the sequences are compartmentalized in the genome (Fischer et al 2004).

Although a majority of minisatellite loci are autosomal, sex-linked VNTR fragments have been found in birds (Rabenold et al. 1990, Millar et al. 1992) and mammals (Ellegren et al. 1994; Kashi et al. 1990). Domingo-Roura et al. (1997) observed in *Felis rufus* (Felidae) the existence of chromosomal bands exclusively found in males, suggesting that these sex-linked bands are Y chromosome specific and are located in the non-recombining region of the Y chromosome, and these repeats probably do not represent conserved chromosomal region.

Several studies have found a relative abundance of microsatellites among mammals, invertebrates and plants, and certain repeat motifs are more common than others. In human, $(A)_n$ and $(CA)_n$ are by far the most common motif variants, the latter being the most widely studied marker. Although a similar situation exists in other mammals (Beckmann and Weber 1992), the same pattern is not true for all taxa. In at least some insect species, $(CT)_n$ is more common than $(CA)_n$ (Estoup et al. 1993), as is the case for an oyster species *Ostrea edulis* (Naciri et al. 1995).

The cytological localization of microsatellites can allow identifying genes under the selection or screening for the genomic reorganization. In *Drosophila subobscura* (Diptera) polytene chromosomes, for example, 72 microsatellite loci with unique hybridization signals were localized. In the sex chromosome, microsatellite distribution was not uniform and its density was higher than observed in autosomes, suggesting convergent evolution for high microsatellite density in the distal part of the X chromosome. The ancestral karyotype within the *Drosophila* genus consists of one dot and five acrocentric chromosome pairs, and there is extensive gene reshuffling within elements, referred to as Muller's elements (Muller 1940). Muller (1940) proposed that the genomes of *Drosophila* species were subdivided into a set of homologous elements represented by chromosome arms, being that each of the five major chromosome arms plus the dot chromosome are given a letter designation (A-F) and that this nomenclature is used to identify equivalent linkage groups within the genus (Schaeffer et al. 2008). The comparative analysis of 72 different types of microsatellites loci among the genomes of *D. subobscura*, *D. melanogaster* and *D. pseudoobscura* allowed deduct the conservation of Muller's chromosomal elements among *Drosophila* species and the intrachromosomal rearrangements within each evolutionary lineage (Drosophila 12 Genomes Consortium 2007, Santos et al. 2010).

The genome of the Hymenoptera species shows high microsatellite abundance compared to other insects. *Nasonia vitripennis*, for example, has approximately 1% of the genome compound of microsatellites. The abundance of microsatellite in *N. vitripennis* shows densities ranging by more than an order of magnitude, both within and among the five chromosomes. In chromosome 1, for example, the density of repeats goes from fewer than 13kb per Mb to more than 30kb per Mb in fewer than 5cM. The microsatellites are compound of dinucleotide motifs, making up 89% of all microsatellites in the three *Nasonia* genomes, while in the genome of *A. Mellifera*, the dinucleotides are also the most abundant motif. Most species of non-Hymenoptera arthropods have an over-representation of a single motif (Pannebakker et al. 2010). The haplodiploid mode of reproduction (Cook 1993, for details see Chapter VIII) observed in hymenopterans can explain the observed pattern of microsatellite diversity and abundance.

The distribution of microsatellites in birds, for example, revealed a density of microsatellites much lower than that estimated for the human genome, which can be explained by the fact that the avian genome contains relatively less noncoding DNA than most mammals. In the same way, avian SINE/LINE (Short interspersed elements/Long interspersed elements) do not terminate in poly(A) tails, which are known to provide a resource for the evolution of simple repeats in mammals. The chromosomal data evidences a dense distribution of $(CA)_n$ repeats over chicken macrochromosomes and intermediate chromosomes, whereas the microchromosomes, a large part of the Z and W chromosomes, and most telomeres and centromeres had very low concentrations of $(CA)_n$ microsatellites. The scarcity of microsatellites on the microchromosomes is compatible to these regions likely being unusually rich in coding sequences (Primmer et al. 1997).

In mammals, for example, the *Sorex araneus* group (Soricidae, Mammalia) is a model for population genetics and evolutionary studies due to the high rate of karyotypic evolution. Basset et al. (2006) mapped 46 microsatellite markers at the chromosome arm level of the *Sorex araneus* group. The results showed evidence for a high degree of conservation in genome organization in the *S. araneus* group despite the presence of numerous Robertsonian rearrangements, and enabling the study of the role of chromosomal rearrangements in the genetic diversification and speciation process of this group.

Sequence analysis demonstrated that the telomere consists of microsatellites repeats (usually of the sequence TTGGGG, or sequences related), and the abundance of microsatellite into telomeres was observed in de protozoa a mamíferos, being conservative for some groups but extremely variable among others (Katzen et al. 1981, Blackburn and Challoner 1984, Ponzi et al. 1985, Baroin et al. 1987, Sugawara and Szostak 1986, Forney et al. 1987, Richards and Ausubel 1988, Allshire et al. 1989). Cytological studies using microsatellites are usually directed to the analysis of telomeric sequences, being useful tools for the establishment of primary chromosomal rearrangements, related to the diploid number reduction or generation of new sex chromosomes (Meyne et al. 1990, Okazaki et al. 1993,

Wan et al. 1999, Nomoto et al. 2001). A good example for these studies is the genus *Muntiacus* (Cervidae), for which the reduction chromosomal observed in *M. vaginalis* (2n=6/7) can be derived from *M. reevesi* (2n=46) hypothetical ancestral karyotype by tandem and centromeric fusions (Hsu et al. 1975). The evolutionary reduction in chromosomal numbers has been studied in detail in this family, and evidence has been found to chromosomal fusion with the use of telomere sequences (Yang et al. 1997a, b).

5.4. TRANSPOSABLE ELEMENTS

5.4.1. General Features of Transposable Elements

Transposable elements (TEs) were discovered by Barbara McClintock several years before the elucidation of the biochemical structure of DNA. TEs are differentiated from other genome sequences by having the ability to mobilize in the genome (Kazazian 2004) and are considered one of the most important mechanisms that drive the evolution of the genome (see Chapter II). The classification of transposable elements is based on enzymology, structural similarities and sequence relationships (Kapitonov and Jurka 2001, Wicker et al. 2007). It includes classes, subclasses, orders, superfamilies, families and subfamilies (Figure 5.6). All eukaryotic TEs belong to two types (retrotransposons and DNA transposons) and are composed of five major classes: (i) long terminal repeat (LTR) retrotransposons, (ii) non-LTR retrotransposons, (iii) cut-and-paste DNA transposons, (iv) rolling-circle DNA transposons and (v) self-synthesizing DNA transposons (Figure 5.7). Each class of TE is composed of a small number of superfamilies or clades, and each superfamily consists of numerous families of TEs (Charlesworth et al. 1994, Hua-Van et al. 2005, Wicker et al. 2007, Kapitonov and Jurka 2008, Pritham 2009). In each family, autonomous elements, which are able to catalyze their own transposition, and nonautonomous elements, which rely on autonomous elements in order to transpose, are found (Richard et al. 2008).

Their number and proportion are highly variable among eukaryotic genomes. The human genome contains about 300,000 copies of DNA transposons, 100 times more than the *C. elegans* genome (Duret et al. 2000, Marin et al. 1998) and 700 times more than the *D. melanogaster* genome (Kaminker et al. 2002). LINEs are non-LTR class I elements whose best characterized member is the LINE-1 (L1) mammalian retrotransposon. In the human genome, approximately 850,000 LINEs were found (Lander et al. 2001), and 660,000 were found in the mouse genome (Waterston et al. 2002); both situations represent approximately 20% of their respective genome size. By comparison, the genome of another vertebrate, *Tetraodon nigroviridis*, contains 700 times fewer transposable elements, almost half of them being LINEs (Jaillon et al. 2004).

The mechanisms for implementation are related to the means used by transposable elements to insert into a new site within the genome. The means of implementation through DNA can be conservative or replicative. First, the TE is removed from one site and inserted into another, while in the second, the TE is duplicated before being transported to a new location, increasing the number of insertions in the genome (Kappitonov and Jurka 2001, Wicker et al. 2007). In the process of transposition via RNA, the RNA intermediate is reverse transcribed into a new copy of DNA, and thus basically replicating, and inserted in a new genomic location (Xiong and Eickbush 1990, Wicker et al. 2007).

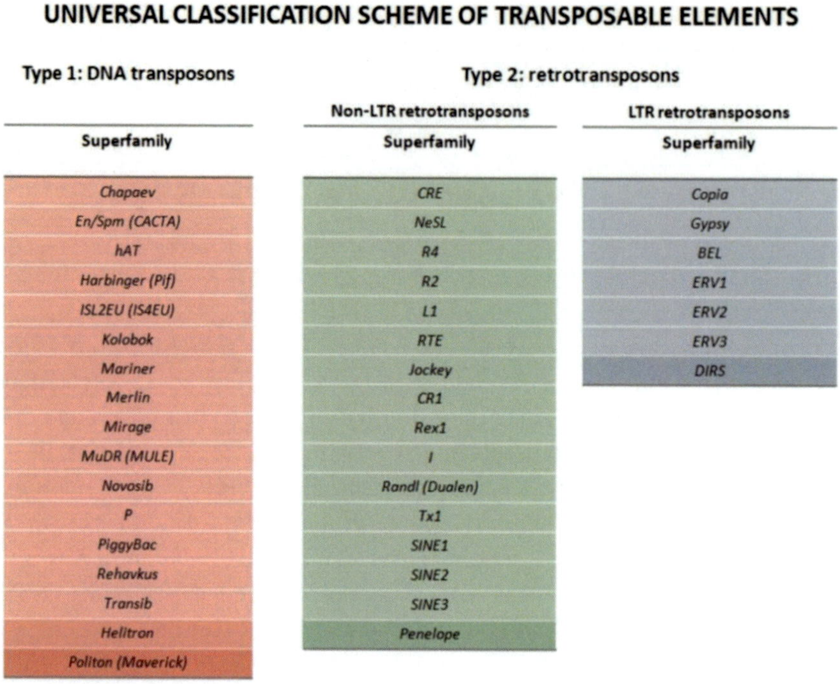

UNIVERSAL CLASSIFICATION SCHEME OF TRANSPOSABLE ELEMENTS

Type 1: DNA transposons	Type 2: retrotransposons	
	Non-LTR retrotransposons	LTR retrotransposons
Superfamily	Superfamily	Superfamily
Chapaev	CRE	Copia
En/Spm (CACTA)	NeSL	Gypsy
hAT	R4	BEL
Harbinger (Pif)	R2	ERV1
ISL2EU (IS4EU)	L1	ERV2
Kolobok	RTE	ERV3
Mariner	Jockey	DIRS
Merlin	CR1	
Mirage	Rex1	
MuDR (MULE)	I	
Novosib	Randl (Dualen)	
P	Tx1	
PiggyBac	SINE1	
Rehavkus	SINE2	
Transib	SINE3	
Helitron	Penelope	
Politon (Maverick)		

Figure 5.6. The universal classification and nomenclature of eukaryotic transposable elements. Different classes of transposable elements (TEs) are differently colored.

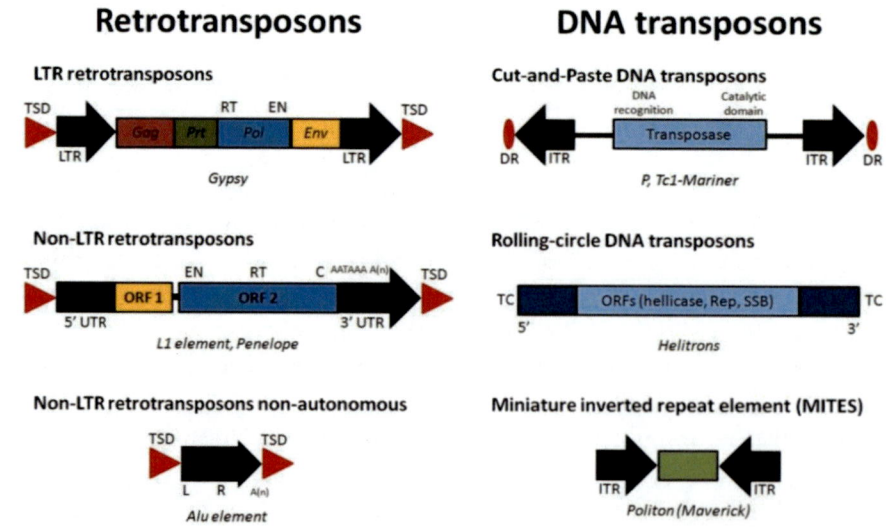

Figure 5.7. Generalized structures of the main types of transposable elements.

5.4.2. Cytogenetic Mapping of Transposable Elements

The transposable elements (TEs) are widely mapped in many karyotypes and in general, their pattern of distribution in heterochromatin and euchromatin is variable among different genomes. However, it seems that they tend to accumulate in the centromeric and/or heterochromatic regions of animal chromosomes. Such patterns can be correlated with a role of the repeated sequences in the structure and organization of pericentromeric regions and heterochromatic areas. However, the distribution of TEs in human and mouse genomes seems to be relatively uniform in both euchromatin and heterochromatin (Kidwell 2002, Volff et al. 2003, Fischer et al. 2004). An astonishing diversity of transposable elements not found in the human and the mouse has been observed in the genome of others animals.

Placozoans are possibly the simplest free-living multicelluar animals and it has been suggested that the placozoan *Trichoplax adhaerens* is an excellent model due its small compact genome and basal position in the eukaryotic phylogeny. DNA transposons were the most abundant TEs in the *Trichoplax* genome, and included diverse superfamilies such as *helitron, piggyBac, hAT, Mariner/Tc1, polinton;* however, TEs accounted for only 0.13% of the *Trichoplax* genome (Wang et al. 2010), which is much lower than the TE content of other genomes. The scarcity of TEs may explain why this genome has undergone the fewest rearrangement rate among metazoan genomes (Srivastava et al. 2008).

In *Drosophila*, TEs accumulate preferentially near centromeres and telomeres and account for 8% of heterochromatin and 4–5% of euchromatin (Bartolomé et al. 2002). Pimpinelli et al. (1995) show the distribution of 9 different transposable elements on *Drosophila melanogaster* mitotic chromosomes (*Copia, Gypsy, mdg-1, blood, Doc, I, F, G* and *Bari-1*) and they are preferentially clustered into one or more discrete heterochromatic regions in chromosomes. Moreover, FISH analysis of geographically distant strains revealed that the locations of these heterochromatic transposable element clusters are highly conserved.

Telomeres in *Drosophila melanogaster* subgroup species are composed of *Het-A, TART,* and *TAHRE* retrotransposons (Villasante et al. 2008). Telomere transposons in lower dipterans could be attractive for the study of diversification of the telomere structure in Diptera and might also provide insights into the evolutionary origin of telomere maintenance based on mobile elements. Another example for maintenance of telomere in insects is the case of *Bombix mori*, in which the telomeric structure is maintained by the transposition of the telomeric retroelements *SART* and *TRAS*. In the silkworm, the genome has telomeric TTAGG repeats and telomerase, but this telomerase shows little or no activity, and the transcription of the telomeric retroelements *SART1* and *TRAS1* is suppressed by *Piwi* subfamily proteins *BmAgo3* and *Siwi*, involved in regulating the transposition of telomeric retroelements (Tatsuke et al. 2010).

The prevalence of non-LTR retrotransposons makes lepidopteran genomes different from that of *Drosophila melanogaster*, in which LTR retrotransposons are the most abundant (Clark et al. 2007). Both the rather high repeat coverage and the prevalence of non-LTR retrotransposons make lepidopteran genomes look like mammalian genomes (Belancio et al. 2008, Cordaux 2008). In *Chironomus piger,* Strenzke (Diptera, Chironomidae) has observed that the genome is more actively rearranging than supposed, due to the presence of retrotransposons *NLRCth1* and *CTRT1*. These retrotransposons are present only in a few loci, but highly variant among different individuals, and a third of breakpoint locations were close to or coincide with a locus occupied by a retroelement (Michailova et al. 2009).

A high number of copies of the rRNA genes are necessary to make the ribosomes needed for protein synthesis. This way, most organisms seem to maintain a surplus of genes, making possible several non-long terminal repeat (LTR) retrotransposons to use these genes as a niche within the genome (Burke et al. 2003, Kojima et al. 2006). The first elements observed within these "rDNA arrays" were the R1 and R2 (R = rRNA genes) elements of *Drosophila melanogaster* that insert into a conserved region of the 28S gene (Dawid and Rebbert 1981, Roiha et al. 1981, reviewed in Eickbush and Eickbush 2007). The R1 and R2 elements have been identified in every lineage of arthropods (Jakubczak et al. 1991, Burke et al. 1998), and more recently recently R2 elements have been found in vertebrates and tunicates (Kojima and Fujiwara 2005), and the R3 element was found in a few insect lineages (Kerrebrock et al. 1989, Burke et al. 1993), the R4 elements were discovered to in nematodes (Neuhaus et al. 1987, Burke et al. 1995) and R5 elements to in platyhelminthes (Burke et al. 2003). The rDNA-R elements can be interpreted as a reciprocal advantage, new inserts for R2 allow a R inserts new elements allow a homogenization of rDNA units, but still do not know if it brings another advantage to the host, although in an evolutionary perspective, this would result in more opportunities for natural selection to operate on the host (Mingazzini et al. 2010).

In basal vertebrates, like fish, a lot of transposons and retrotransposons have been identified (Volff et al. 2003); many of them are also present in invertebrates and most probably were acquired by horizontal transfer. In medaka (*Oryzias latipes*) fish, 7% of its genome consists of transposable elements (Kasahara et al. 2007). The genome of models species, *Takifugu rubripes* and *Tetraodon nigroviridis* contain about 3-4% of DNA transposons (Aparício et al. 2002, Jaillon et al. 2004). Analysis of the chromosomal location of various types of TEs in the compact genome of the pufferfish *Tetraodon nigroviridis* showed that these sequences are generally excluded from gene-rich regions (Dasilva et al. 2002). Among cichlids, some retrotransposable elements (*Rex1*, *Rex3*, *Rex6*) were compartmentalized predominantly in the centromeric region in coincidence with heterochromatic areas and also as small dispersed signals along most chromosomes (Teixeira et al. 2009, Mazzuchelli and Martins 2009, Valente et al. 2010). The compartmentalization of *Rex1* and *Rex3* elements in the centromeric areas and heterochromatin was also observed in the fish *Notothenia coriiceps* (Ozouf-Costaz et al. 2004).

The centromeres of a number of species have also shown TEs interspersed with satellite sequences (Hua-Van et al. 2005). Fischer et al. (2004) showed a heterochromatic co-localization of transposable elements with minisatellites in the compact genome of the pufferfish *T. nigroviridis*. The satellite DNA amplification together with retrotransposon accumulation in the centromeres indicates an important role for the expansion and stabilization of this chromosome region (Plohl et al. 2008).

Tol2 transposable element was identified from the genome of the medaka fish (*Oryzias latipes*), a small fresh water teleost, and is similar to those transposons belonging to the hAT family, namely *hobo*, *Ac* and *Tam3*. *Tol2* can integrate almost everywhere in the genome, and there are no preferences with respect to position on a chromosome (proximal or distal to the telomere or the centromere). *Tol2* can transpose in various vertebrates, indicating that, although host factors involved in the process of transposition of *Tol2* have not been identified, such factors should be conservatively present in vertebrates (Kawakami 2007).

At least 12 classes of retrotransposons identified in fish are also present in amphibians (Volff et al. 2003) More than one-third of the frog genome consists of TEs, higher than the 9% TE density in the chicken genome but comparable to the 40 to 50% density in mammalian genomes. Many families of frog TEs are more than 25% divergent from their consensus sequence, so like mammalian and bird TEs that have persisted for as long as 20 to 200 million years. This contrasts with the faster turnover observed in insects, nematodes, fungi, and plants (Kapitonov and Jurka 1999, 2003). In contrast to those of other vertebrates, the most recognizable frog TEs (72%) is DNA transposons, rather than the retrotransposons that dominate other genomes. The frog's genome also contains LTR retrotransposons of all major superfamilies, with higher diversity than in all other studied eukaryotes. Although most are ubiquitous, *Copia*, *BEL*, and *Gypsy* elements are not found in birds and mammals, suggesting that this subset became immobile after divergence from the amphibian lineage (International Chicken Genome Sequencing Consortium 2004; International Human Genome Sequencing Consortium 2001; Mouse Genome Sequencing Consortium 2002).

Tc1-like transposon belongs to the superfamily *Tc1/Mariner* and is exceptionally widespread in living organisms, ranging from protozoa to vertebrates (Doak et al. 1994). A sequence that originated from a *Tc1*-like element was organized in tandem arrays in the centromeric area of few chromosomes in *Rana sculenta* (frog) (Pontecorvo et al. 2000). In fish, chromosomal mapping data on *Tc1*-like sequence in the cichlid *C. kelberi* revealed this transposable element in centromeric regions and dispersed along the chromosomal arms of most chromosomes (Teixeira et al. 2009). Another example of such remarkable compartmentalization was observed in *T. nigroviridis*, where the *Tc1*-like elements were clustered in the heterochromatic short arms of six subtelocentric chromosome pairs (Dasilva et al. 2002).The success of the *Tc1* family throughout the animal kingdom provides an excellent opportunity to examine the evolution of these elements in closely related host species as well as in distant phyla. Horizontal transmission has been documented as a major mechanism for the dissemination of *Tc1-Mariner* element among species. It has been shown that a *Tc1*-like transposon from a nematode jumps in fish cells and that a fish *Tc1*-like transposon jumps in human cells (Ivics et al. 1997).

Sauropsida is a group of amniotes that includes all known reptiles, dinosaurs, and birds. Research on TEs of sauropsids was initially concentrated on birds, the first TE to be characterized being chicken repeat 1 (*CR1* LINE) (Stumph et al. 1981, 1984, Silva and Burch 1989). Later, in the mid 1980s, the first investigations of SINEs in sauropsids started with the analysis of SINEs in turtles (Endoh and Okada 1986, Endoh et al. 1990). The first evidence for the presence of Ty/*Copia* elements in squamate genomes and the first insight into the presence of Ty3/*Gypsy* elements in reptilian genomes was provided by PCR analysis in 1990s (Flavell et al. 1995, Miller et al. 1999).

The TEs most intensively investigated in Sauropsida have been the non-LTR retrotransposons (LINEs or retroposons), such as RTE clade (including *Bov-B* LINEs) (Gogolevsky et al. 2008, Novick et al. 2009, Piskurek et al. 2009), *L2* clade, *L1* clade (Piskurek et al. 2009), *R4* clade (Novick et al. 2009, Piskurek et al. 2009), *CR1* clade (Novick et al. 2009, Piskurek et al. 2009, Shan et al. 2009) and *R2* clade (Kojima and Fujiwara 2005). Chromosomal distribution of TEs has been analyzed so far only in the chicken genome and has not been studied in any reptilian genome (Kordis et al. 2010). Until the availability of the first draft of the *Anolis* (lizards) genome, it was not clear whether reptiles possess more mammalian-like or more fish-like TEs and only a few site-specific repetitive sequences have been isolated and characterized in reptiles (Singh et al. 1980, Panicker and Singh 1994, Capriglione et al. 1998). Indeed, the *Anolis* genome provided the first direct evidence for more fish-like repetitive landscape and TEs of reptiles (Novick et al. 2009, Piskurek et al. 2009). Although the *Anolis* genome provided an excellent insight into the repetitive genome landscape, TEs and the diversity of TEs in lepidosaurs, we still have very limited genome sequence data for turtles and crocodiles.

In chicken, genome sequencing revealed a lower density of interspersed repeat elements (comprising about 10% of the genome) than is typically seen in mammals (40–50%), and the limited variation in DNA content observed across birds suggests that avian genomes, in general, are low in repeats. The evolutionary stability of bird chromosomes might relate to a low frequency of transposable elements (Ellegren 2010). The percentage of a chromosome occupied by repeats in chicken correlates positively with chromosome length, but even the largest chicken chromosome has repeat densities much lower than those in mammals, indicating that avoidance of repeats in the chicken is not confined to microchromosomes (Hughes and Piontkivska 2005). The distribution of relatively young CR1 families is enriched in regions of high GC content and is more abundant on the Z sex chromosome than on autosomes, probably due to its low recombination rates (Abrusan et al. 2008). The *Mariner* transposons that are present in the genomes of fish, amphibian and primates, reaches high copy numbers in the chicken genome (International Chicken Genome Sequencing Consortium 2004, Wicker et al. 2005), but it is not present in any other available avian genome data (emu, condor, turkey and zebra finch). This clearly indicates that *Mariner* jumped into the chicken genome by horizontal transfer (cited in Kordis et al. 2010).

SINEs are the most abundant elements in mammalian genomes. They include *Alu* and MIR elements in primates and *B1*, *B2*, and *ID* elements in rodents as well as many other elements in mammalian and non- mammalian genomes. Both the human and the mouse genomes contain approximately 1,500,000 SINEs, which make them the most abundantly repeated elements in these genomes (Lander et al. 2001, Waterston et al. 2002).

FISH mapping of *Alu* elements to human metaphase chromosomes demonstrated that these elements exist in R bands, a distribution similar to that of the unrelated *B2* SINE distribution in the mouse genome (Boyle et al. 1990). The distribution pattern observed for the rat *ID* elements is similar to those of human *Alu* elements and mouse *B2* SINES, where all were localized to the R bands (Ono et al. 2001). However, FISH mapping of SINEs (not necessarily the major class) in the ovine and bovine genomes demonstrated a diffuse hybridization pattern and an absence from the R bands (Rajcan-Separovic and Sabour 1993).

Chromosomal mapping of the Mon core-SINEs to platypus chromosomes demonstrated a high concentration on the distal regions of the six large autosomal pairs and the largest of the five X chromosomes. Mon core-SINEs hybridized to euchromatic regions and was absent from heterochromatic regions (Kirby et al. 2007). The distribution of signals on platypus chromosomes bears a striking resemblance to the R-band pattern of platypus chromosomes reported by Wrigley and Graves (1988). This is consistent with observations of the distribution of SINEs in R-banded regions of the human and mouse genome. Concentration of repetitive sequences in the distal regions is an unusual pattern, for distal chromosomal regions are normally GC-rich and gene-rich in warm blooded animals (Perani et al. 2000).

LINEs are one of the major components of eukaryotic genomes (Kazazian 2004). They are classified into 5 major groups (*L1*, RTE, *Jockey*, *R2* and *I*), each containing numerous clades (Malik et al. 1999). *L1*s are abundant retrotransposons that comprise about 20% of mammalian genomes. The chromosomal distribution of the interspersed-type repetitive sequences has been described in the species of the mammal's genus *Microtus* (short-tailed vole). The *L1* and non-*L1* retroposons are interspersed in all autosomes and are especially abundant in the whole heterochromatin regions of the X and Y chromosomes (Neitzel et al. 1998, 2002). In addition to their presence in the sex chromosomes, the two retroposon sequences were amplified in the heterochromatic arms of autosomes (Yamada et al. 2006).

In an overview, we can see that mammal's transposable elements characterized to date appear to be non-randomly distributed. While most TEs were found to be exclusively restricted to specific chromosomal location, LINE elements and some retrovirus-like elements are preferentially accumulated in G-banding regions of the chromosomes, and in some cases, in the sex chromosomes, while SINE elements occur preferentially in R-banding regions. Four mechanisms are presented which may explain the non-random genomic distribution of mammalian transposons: i) sequence-specific insertion, ii) S-phase insertion, iii) ectopic excision, and iv) recombinational editing. Some of the available data is consistent with each of these four models, but no single model is sufficient to explain all of the existing data (Wichman et al. 1992).

5.4.3. Conclusion Remarks on Cytogenetic Mapping of Transposable Elements

Besides the preferential distribution of TEs in non-coding regions, they have a different distribution among and within chromosomes, being most frequently associated with sex chromosomes than autosomes, probably due to a higher concentration of heterochromatin in these elements (Kidwell 2002). These sequences have an important role in sex chromosome differentiation and evolution. Together this data indicates that transposable elements are major structural components of heterochromatin and have played an important role in the chromosomal evolutionary history of host genomes.

It is believed that some factors such as gene density, chromatin structure and recombination rate may have a role in the maintenance of TEs at a specific genomic region. Generally, TEs seem to be more abundant in heterochromatin in several genomes, and their presence in these regions seems to be common among multicellular eukaryotes. Despite some discordance, there are many possible explanations for the relationship between TEs and heterochromatin: (i) TEs tend to accumulate in regions with low recombination rates as a consequence of their removal from regions with high recombination rates, where ectopic recombination could have more deleterious effects; (ii) there is more elimination of TEs in gene-rich regions because of their potential deleterious effects when inserted within genes; (iii) the high expression of TE-encoded products could have negative consequences for the genome due the cost for the cell; thus, these TEs would be eliminated from regions with high expression levels; (iv) TEs could accumulate in heterochromatin as a consequence of their functional involvement in the maintenance of specific genomic regions, such as the pericentromeric and telomeric regions.

Special attention must be exercised concerning the relationship between the data obtained through the cytogenetic mapping of dispersed repeated DNAs and the data provided through the complete sequencing of genomes. The dispersed signals of transposable elements revealed through the molecular cytogenetic analysis does not represent dispersed single copies, but small clusters of at least a few copies of the DNA element. The current FISH procedure conducted in most cytogenetic laboratories permits the visualization of segments of DNA that encompass at least 10 kb. In this way, several copies of a transposon have to be closely organized in the genome to generate a single detectable signal onto chromosomes. At the same time, dispersed cytogenetic signals need careful analysis concerning several aspects of the FISH procedure employed as size, amount and nucleotide content of the probe, and stringency conditions of the hybridization procedure. Such FISH conditions can generate background easily misunderstood as real signals in the chromosomes.

Chapter 6

MAPPING OF SINGLE-COPY GENES

Recently, many studies have been conducted using dispersed or *in tandem* repetitive DNAs as probes for FISH cytogenetic mapping. In general, these probes provide easily visible signals, due to their abundant repetition and distribution pattern forming long blocks along the chromosome (see the topic "Mapping of repetitive DNAs" in this chapter). However, to determine small chromosomal rearrangements such as translocations and inversions or even to determine an exact position of small DNA sequence like single-copy genes, a more refined methodology based on single copy fluorescent *in situ* hybridization is required. It is now straightforward to combine methods that range from identifying gross structural changes in the genome down to single-nucleotide differences. Various technical modifications and in-house tricks have been developed by different laboratories to improve FISH sensitivity. Such modifications involve improvement in specificity and resolution of the FISH technique, brought about by a better understanding of the chemical and physical properties of nucleic acids and chromatin, together with the advances in the fields of fluorescence microscopy and digital imaging, associated to the growing availability of genomic and bioinformatics resources.

But how can a small probe be visualized with FISH? Unfortunately, the usefulness of FISH in specific applications is still frequently limited by low detection sensitivity. For example, using current FISH techniques, a probe containing less than 10 Kb of single-copy DNA did not reliably reproduce detectable signals in metaphase chromosomes. To overcome difficulties in the chromosome localization of single-locus genes, two basic and different approaches have been developed. One is the use of genomic DNA clones, like large cosmid, yeast artificial chromosome (YAC) or bacterial artificial chromosome (BAC) clones, containing single-copy genes to be mapped, as probes for FISH. Another one is a sequence-specific detection of nucleic acid *in situ* that did not require large and labeled hybridizing probes. This method is called primed *in situ* labeling (PRINS) and is based on rapid annealing of unlabeled DNA primers to complementary target sequences, followed by *in situ* primer elongation with *Taq* polymerase.

Besides, there is the possibility of combining these technologies on the extended chromosomes present in interphasic or meiotic chromosomes. It has become a usual practice as an alternative to mitotic chromosomes to map single or low copy sequences to resolve distances between sequences smaller than 1 Mb. Moreover, there is the fiber-FISH technology that is one of the most powerful tools for mapping small DNA sequences onto specific regions of the genome, because it allows accurate sizing of gaps and overlaps between probes. These techniques are called high-resolution FISH and represent an alternative to achieve information on genetic collinearity that allows a better understanding of the processes that are involved in normal genomic function, diseases, and chromosomal evolution and rearrangements.

6.1. STRATEGIES FOR MAPPING SINGLE-COPY GENES

Various types of genomic DNA clones have been used in FISH mapping, including λ clones (Peterson et al. 1999), cosmid clones (Sadder and Weber 2002), YAC clones (Fransz et al. 2000) and BAC clones (Tör et al. 2002) that are powerful tools for detailed analysis of complex genomes. Among them, the most recently used for cytogenetic mapping are the BACs, that, although presents insert sizes (typically 100~300 kb) much smaller than those of YACs, have several additional advantages over YACs such as free from chimerism, higher stability of the insert and are very easily manipulated (Hasterok et al. 2006). Due this statement,most parts of the studies have been conducted using BACs as probes as will be discussed below.

BACs have been demonstrated as useful for many aspects of molecular and genomic studies, such as the positional cloning of genes (Stiglec et al. 2007a), comparative studies of synteny and gene organization among different species (Goldammer et al. 2009), as well as for local or whole genome physical and genetic mapping and sequencing (Katagiri et al. 2005). In cytogenetic research and chromosome mapping, BAC clones have also been used successfully as probes in FISH (Jiang et al. 1995, Hoskins et al. 2000, Stiglec et al. 2007a). *In situ* hybridization of these sequences allows recognizing each chromosome pair individually, integration of physical and genetic maps, and determining the chromosomal relationships among related species, as well as species belonging to different orders, through cross-species hybridization (Cheng et al. 2001, Pedrosa et al. 2003, Hasterok et al. 2006, Ferguson-Smith and Trifonov 2007, Griffin et al. 2007). The potential of FISH for studying a lot of animal genomes has increased considerably after the construction of a number of commercial and non-commercial BAC libraries: giardia*, trypanossoma*, fruitflies*, lepidopterans (Wu et al. 2009), honey bee*, channel catfish (Quiniou et al. 2003), Atlantic salmon (Thorsen et al. 2005), stickleback*, zebrafish*, Nile tilapia (Katagiri et al., 2005), frog*, reptile green anole lizard*, turkey*, chicken*, mouse*, monkey*, bat*, rat*, dog (Werner et al. 1999), bovine*, caprine* and of course human*. (* indicates commercial BAC libraries available at http://bacpac.chori.org/home.htm).

From those libraries a number of molecular markers have been identified and sequenced and their chromosomal positions have been established. BACs have narrowed the gap between cytogenetics and molecular genetics and have become important tools for visualizing the organization of genomes because they produce bright, well-defined signals on metaphase and interphase chromosome preparations (Figure 4.2). BACs constitute a resource that provides the tools to integrate questions raised at the chromosomal level into answers at the molecular level (Korenberg et al. 1999).

There are some examples in the literature showing the power of BAC-FISH analysis. Korenberg et al. (1999) show how BACs provide molecular links for understanding human genomic duplications, meiosis, and evolution, as well as application for conducting genome-wide prenatal diagnosis at the molecular level and for detecting candidate genes associated with novel cancer breakpoints. Like in humans, a lot of experiments have been done using BACs as probes in animal genetics: Corradini et al. (2003) have performed FISH mapping of nine bacterial artificial chromosomes (BACs) carrying several predicted genes to correspond to protein-coding genes involved in important cellular functions in *Drosophila*. Experiments have been done in chicken, where 17 genes were mapped through BAC probes onto female chicken metaphase spreads (Stiglec et al. 2007a).

In the pufferfish (*Fugu*) and in the Antarctic fish (*Notothenia coriiceps*) it was demonstrated through BAC-FISH that two of the globin gene locus were located on separate chromosomes (Gillemans et al. 2003, Pisano et al. 2003). Likewise, Harvey et al. (2003a) and Chiang et al. (2001) have demonstrated that the two aromatase genes (CYP19), which catalyses the conversion of androgens are located in distinct chromosomes in the Nilo tilapia *Oreochromis niloticus*, and zebrafish (*Danio rerio*).

The use of BAC-FISH on comparative cross-species hybridization has increased in the last years in animal cytogenetic research. A detailed comparative analysis of sheep, human, and cattle mapping data allowed the construction of a comparative map that confirms and expands the knowledge about evolutionary conservation and break points between the X chromosomes of the three mammalian species (Goldammer et al. 2009). In lepidopterans, recent advances in *Bombyx mori* genomics make comparative genomic studies feasible. The draft sequence information of this species has been deposited in a public database and all genetic linkage groups have been successfully assigned to individual chromosomes (Yoshido et al. 2005). BAC-FISH mapping using selected BACs carrying orthologs genes are powerful tools for the identification of conserved synteny between *B. mori* chromosomes and the chromosomes of other lepidopteran species (Sahara et al. 2007).

Comparative cytogenetics based on BAC-FISH between Atlantic salmon and rainbow trout was able to anchor genetic maps (linkage map) to chromosomes (physical map) (Phillips et al. 2009). This analysis provided strong evidence for conservation of large syntenic blocks in these species, corresponding to entire chromosome arms in rainbow trout. BACs containing sex linked markers of the Nile tilapia were mapped in the chromosomes and compared to the linkage map of the species (Cnaani et al. 2008). Another importance of these assays is related to the possibility of mapping the real distances between genes and marks in the chromosomes. This information cannot be predicted using linkage maps, because recombination rates vary along the length of chromosomes according to genomic content and, as a result, genetic distances are not directly proportional to physical distances.

In spite of the power of BAC mapping, one should be aware that these large-insert genomic clones as BACs, especially those from species with very large genomes, may contain an extensive amount of repetitive DNA sequences that could decrease the efficiency of localization of the single-copy sequences and also causes unspecific hybridization (see Figure 4.2). However, if repetitive sequences account for 90% of an animal genome, a 100-kb BAC clone should, theoretically, contain ~10-kb of unique sequence, which is enough to generate FISH signals. The cross hybridization from the repetitive DNA sequences can be minimized by pre-annealing the probe with C_0t-1 DNA or shared genomic DNA (Jiang and Gill 2006).

When an unpredicted location of a BAC clone is identified, additional BAC clones from the same genomic region should be mapped to understand whether observed results were caused by chimeric BAC structure, sequence similarity between different genes, or chromosomal insertions that were not detected previously (Kukekova et al. 2009). A critical step to obtain reliable results is the slide preparation with chromosome spreads and the pre-treatment of the slides for the FISH procedure. Differences in these steps may be the reason why FISH protocols are frequently different for different species.

In an attempt to minimize some problems occurred througgh BACs hybridization due to the small signal generated, PRINS (primed *in situ* labeling) represents a different strategy for single-copy mapping. According to the PRINS method, which combines features of fluorescence *in situ* hybridization (FISH) and polymerase chain reaction (PCR), unlabeled short oligonucleotide primers are annealed and extended on chromosome preparations on microscope slides in the presence of labeled nucleotides. During primer elongation, labeled nucleotides are incorporated into newly synthesized DNA. After extension, the relevant sequences are visualized by fluorescence microscopy (Koch et al. 1991).

The use of oligonucleotide probes allows the detection of much smaller targets than those accessible to conventional *in situ* hybridization and results in the higher sensitivity of the PRINS method and in its amazing discriminatory power for small sequence variations and detection of low and single copy DNA sequences (Pellestor et al. 1994, 1995). The signal intensity can be further increased by performing several cycles of the PRINS reaction (so-called cycling PRINS), which accumulates labeled copies of the target sequence at the site of synthesis (Terkelsen et al. 1993).

The ability to localize single copy genes via PRINS methodology has opened various applications in diagnosis and research. For example, PRINS was used to localize the FACTOR IX gene (a single copy gene on the X chromosome) (Cinti et al. 1993) and in the elucidation of specific genetic defects, such as abnormal sex differentiation through the localization of SRY gene in XX men, in a woman with XY gonadal dysgenesis and in an azoospermic man with Xp-Yp interchange (Kadandale et al. 2000). In addition, PRINS demonstrates efficiency, sensitivity and specificity in the diagnosis of microdeletions in groups of patients with Prader Willi/Angelman (PWS/AS) syndrome and DiGeorge/velocardiofacial syndrome (DGS/VCFS) (Tharapel et al. 2002).

Although there is the possibility of using PRINS for mapping of single copy sequences, this methodology has been mostly applied in mapping repetitive sequences as centromeric and telomeric sequences, *Alu* repeats and 5S rRNA genes (Koch et al. 1991, Gosden et al. 1991, Gosden and Lawson 1994, Martins and Galetti 1999). The PRINS was also applied for the detection of repeated DNA sequences in several domestic animals, such as chicken (Coullin et al. 2005), pig (Pinton et al. 1998), horse (Wnuk et al. 2008) and frog (Freeman and Rayburn 2005).

PRINS proved to be useful because (i) it generates highly specific labeling and the sensitivity and specificity of PRINS allow localization of DNA segments that are too small to be detected by conventional FISH. Moreover, the target-specific PRINS approach can overcome some FISH limitations such as the non-specificity of the usual probes (FISH probes can contain overlapping DNA segments outside the targeted loci and can cross-hybridize to unintended targets); (ii) it can be performed rapidly and easily; PRINS results can be available within 2±4 hours, whereas a typical FISH procedure takes at least 12 hours. (iii) single-copy genes and small DNA segments in general, can be detected simultaneously by combining PRINS and FISH; (iv) it is useful for analysis of chromosome rearrangements such as translocations or small intragenic deletions; (v) this method leads to a considerable reduction of costs approximately 10 times less expensive than FISH (Cinti et al. 2002, Coullin et al. 2002, Tharapel et al. 2002, Wilkens et al. 1997). Although this methodology presents several advantages, it has not been explored in basic research.

In order to achieve the goal of routine single-copy gene detection, in some cases, it is necessary to combine these methods with a completely new technique that would provide better target access into the chromosomal DNA like the use of the extended chromosomes present in meiotic interphase or pachytene.

6.2. IMPROVING SINGLE-COPY MAPPING

In order to improve the resolution of metaphase chromosomes, FISH has been applied to interphase cell nuclei, meiotic chromosomes, mechanically stretched chromosomes (MSCs) and by fiber-FISH (Sallinen et al. 2001). For example, the use of meiotic chromosomes that are considerably longer than their mitotic counterparts can increase to 2 Mb the FISH resolving power (Cheng et al. 2002) and the resolution in the fiber-FISH method ranges from a few to 300 Kb (Volpi and Bridger 2008). This is named high-resolution FISH and includes the hybridization of probes to "free chromatin." These targets have widened the resolution of FISH to detect distances from the traditional cytogenetic resolution level down to a resolution of a few kilobases (Sallinen et al. 2001). This method can help to map single-copy sequences directly on chromosomes and also to order the BAC contigs better solving problems associated with physical mapping from genome projects, e.g., ambiguous gene orders and intergenic distances, concerning the comparison of the human and mouse genomes (Sallinen et al. 2001).

For example, birds, like other vertebrates with yolk-rich eggs, present lampbrush chromosomes during the diplotene stage of the first meiotic prophase (Hutchison 1987) and theses chromosomes can certainly provide excellent resolution to locate repetitive and single-copy sequences on avian chromosomes such as chicken (Hori et al. 1996, Solovei et al. 1998, Derjusheva et al. 2003, Galkina et al. 2005, 2006).

In certain organisms, the use of FISH on pachytene chromosomes has become usual practice to map single or low copy sequences because they allow a higher degree of spatial resolution, and also display a clear differentiation of heterochromatin blocks, which helps the identification of individual chromosomes, rearrangements and translocations, and establishes repeat-rich domains (Pigozzi 2007).

However, the resolving power of pachytene FISH depends on three factors. First, the degree of chromatin condensation at the pachytene stage may vary significantly among different species. Second, the location of probes in euchromatic or heterochromatic regions will affect the FISH resolution. Third, the resolving power also depends on the different pachytene substages (early or late). Late pachytene chromosomes can be used to orient the telomere-centromere positions of the adjacent clones, whereas early pachytene chromosomes can be used to resolve even partially overlapped BAC clones (Cheng et al. 2002). Because of these three factors, special attention should be paid to convert the microscopic distance, which separates the two DNA probes on pachytene chromosomes, into kilobases (Cheng et al. 2001).

Meiotic bivalents prepared for observation of the synaptonemal complexes (SCs) have been employed in plants and also among vertebrates (Peterson et al. 1999, Froenicke et al. 2002, Pigozzi 2007). The physical location of nine genes was established using FISH along the ZZ bivalent, using synaptonemal complex spreads from the avian male Zebra finches and it was possible to compare their positions in the mitotic and meiotic chromosomes (Pigozzi 2008). Single-copy sequences have also been mapped on mouse pachytene bivalents using SC and their positions compared with those obtained by genetic analyses (Froenicke et al. 2002). This kind of comparison seems to be very important due it showing that gene positions with respect to the centromere are different in mitotic compared to meiotic chromosomes in mice (Froenicke et al. 2002).

Similarly, several differences were observed in relative length and arm ratios between mitotic chromosomes and their SCs in human spermatocytes (Sun et al. 2004). These discrepancies lie in the differential packing of the DNA along the axial elements of the highly specialized meiotic chromosomes. The main components involved in this differential distribution of the DNA are repetitive sequences enriched in mammalian G or R bands and those sequences forming heterochromatin (Stack 1984, Zickler and Kleckner 1999, Hernandez-Hernandez et al. 2008).

Among fishes, FISH and DAPI staining of the synaptonemal complex of the Nile tilapia (*Oreochromis niloticus*) allow orientation of the unpaired region of bivalent 1 observed during early pachytene. This appears to be related to recombination suppression around a sex determination locus (Ocalewicz et al. 2009).

Mechanically stretched chromosomes (MSCs), prepared by cytocentrifugation of hypotonically treated cells, provide an alternative target for high-resolution FISH mapping. However, the morphology of MSCs is significantly distorted so that little, if any, banding patterns are discernible. Furthermore, the degree of stretching is variable from one slide to another, from one metaphase to another, and from one chromosome to another. Thus, no quantitative information on precise probe distances can be obtained (Laan et al. 1995).

The application of extended chromosomes for the ordering YAC clones separated by less than 1 Mb and for the structural analysis of α-satellite DNA and proteins, has been reported in humans (Haaf and Ward 1994a, b). Sallinen et al. (2001) have applied both MSCs and extended DNA fibers to the physical mapping of the mouse genome. At first, five mouse collagen genes were localized and the mutual order of the genes, centromere–*Col10a1*–*Col13a1*–*Col6a2*–*Col6a1*–*Col18a1*–telomere, was determined.

The fiber-FISH is another technique also applied to the high-resolution FISH mapping approach of small DNA sequences onto specific regions of the genome and ordering of probes relative to one another can be also performed on released chromatin fibers. Chromatin can be released from interphase cell nuclei by various chemical or mechanical methods, and investigators tried to coin names that reflect their individual approach. Isolation of DNA from cell nuclei, extension, and preparation of chromatin or DNA fibers with diameters ranging in size from a few to several hundred nm improves the accessibility of the DNA targets for both probes and detection reagents (Weier 2001). This method is essential for defining gene order and organization, the physical architecture of specific genomic regions, allows accurate sizing of gaps and overlaps in contigs, and analysis of segmental duplications and copy number variants. The degree of resolution in the fiber-FISH method ranges from a few to 300 Kb (Volpi and Bridger 2008).

The method consists of removing the histones of chromatin and stretching and fixing of DNA on a slide prior to hybridization (Volpi and Bridger 2008). The sensitivity of the technique is increased, probably as a consequence of the high accessibility of naked DNA to probes and immunological detection reagents. Fiber-FISH has been used in various types of animal genome mapping projects, including the analysis of structure and organization of repetitive DNA sequences, mapping of single-copy, large genomic locus, mapping of BACs and comparative analyses.

Theuns et al. (1999) determined the genomic organization of the human presenilin 1 gene, which was localized on chromosome 14q24.3 by fiber–FISH. Genes of the major histocompatibility complex (MHC) of rhesus macaque were localized to the long arm of the rhesus macaque chromosome 6 in 6q24, the orthologous region to human 6p21.3. Furthermore, centromere to telomere orientation of the rhesus macaque MHC as well as the internal order of the MHC genes tested is the same as in humans. Fiber-FISH allows a rough estimate of distances between these MHC genes in the rhesus macaque, and, as in the human, the rhesus macaque MHC comprises about 3 to 4 Mb (Huber et al. 2003).

The Rhesus (Rh) blood group system in humans is encoded by two genes with high sequence homology. However, the genomic organization of Rh genes in chimpanzees and other non-human primates has not been precisely studied. Suto et al. (2003) analyzed the arrangement of the Rh genes of chimpanzees *(Pan troglodytes)* by two-color fluorescence *in situ* hybridization on chromatin DNA fibers (fiber-FISH) and discovered intra- and interspecific genomic variations in the Rh gene locus in hominoids, so it would shed further light on reconstructing the genomic pathways of Rh gene duplication during evolution.

A comparative fiber-FISH on extended chromatin of cattle (*Bos taurus*, BTA, 2n = 60), river buffalo (*Bubalus bubalis*, BBU, 2n = 50), sheep (*Ovis aries*, OAR, 2n = 54) and goat (*Capra hircus*, CHI, 2n = 60), using the SMN gene, was performed on R-banded chromosome preparations. SMN was located on BTA20q13.1, OAR16q13.1, CHI20q13.1 and BBU19q13. These chromosomes and chromosome bands are believed to be homeologous, confirming the high degree of chromosome homeologies among bovids (Iannuzzi et al. 2003).

The fiber-FISH have been used to examine the structural organization of the mouse a2u-globulin locus, and found the genes to be arranged as an array of both direct and inverted repeats. The organization of the rat a2u-globulin genes differs from the mouse genes and suggests different evolutionary events that have acted reorganizing these homologous sets of genes (McFadyen and Locke 2000).

The high resolutions of the pachytene FISH method, together with the recently developed fiber-FISH techniques and MSCs, add new tools to the arsenal for fine physical mapping of single-copy genes, and, thereby, in positional cloning and genomic sequencing of animal species. Consequently, this would open new avenues in the field of animal comparative genomics.

Chapter 7

CHROMOSOME PAINTING IN COMPARATIVE CYTOGENETICS AND GENOMICS

Advances in molecular cytogenetics have provided great insights into chromosome evolution, with the chromosome painting being a powerful technique to achieve this aim. This technology allows the application of whole chromosome (WCP-whole chromosome painting) or a partial chromosome (PCP-partial chromosome painting), that may be obtained by microdissection or flow-sorting methods followed by an amplification and labeling step, as probes. Usually, this technique has been applied to the study of the evolution of whole karyotypes, or particular chromosomes, such as B chromosomes, specific chromosomes from A complement, and sex chromosomes. It is common to use these probes when it is necessary to identify the homologies or homeologies among some chromosome -or all chromosomes- from distinct species. This kind of hybridization is named cross-species chromosome painting, or comparative chromosome painting or Zoo-FISH (revised by Yang and Graphodatsky 2009). The reciprocal cross-species painting allows the more exact identification of homologous chromosome segments. However, the Zoo-FISH is not successful among groups that diverged genetically more than 105 million years (Ferguson-Smith and Trifonov 2007).

The first use of chromosome painting in vertebrates was performed in a human in 1988 by Pinkel et al. (1988) and Lichter et al. (1988), and today it is very common to see the application of this technique in other mammals as well as birds, fishes and some invertebrates. One of the first applications for comparative cytogenetic was conducted by Wienberg and colleagues (1990), who applied this tool for comparative analysis between human and Japanese macaque genomes. Thereafter, a great number of studies have allowed the comparison among genomes, mostly of eutherian mammals and birds (Wienberg et al. 1990, Jauch et al. 1992, Shetty et al. 1999, revised by Griffin et al. 2007, revised by Ferguson-Smith and Trifonov 2007, Stanyon et al. 2008, revised by Yang and Graphodatsky 2009, among others).

In the beginning, the chromosome paintings were obtained from libraries of human chromosome-specific DNAs cloned in BACs (revised by Yang and Graphodatsky 2009). Each BAC clone may be used to physical mapping and sometimes one or more BAC clones may be able to paint one chromosome completely or partially. BACs obtained from one species can be used to screen the genomic library of another species and thus, the homologous BAC clones of the latter species may be used to its own physical mapping, showing the regions of homology with the first species. This comparative BAC mapping is a kind of "cross-species chromosome painting" and this methodology is still used today when the species that are under study have high chromosomal DNA divergence such as the eutherian and monotremes or the marsupials and the other vertebrate classes (revised by Ferguson-Smith and Trifonov 2007).

The most common studies about chromosome painting were conducted in mammals because of the difficulty in the probe achievement for other animal groups (see Chapter IV). The achievement of sex chromosomes, B chromosomes, and other chromosome markers as probes are facilitated if the length, shape or heterochromatin patterns are different in relation to the other chromosomes from their own genomes. Nevertheless, there are many papers in the literature that conducted chromosome painting in other groups such as invertebrates, birds, fishes, among others. It will be revised in this section data concerning chromosome painting for genomic comparison studies in a range of taxonomic groups.

Before the advent of chromosome painting in mammal groups, the comparison among karyotypes was conducted based on classical cytogenetic methods (for example GTG-banding technique) as well as mapping of individual genes or other kind of DNA sequences. After the advent of chromosome painting, the knowledge acquired in the field of classical cytogenetics were confirmed and refined. In birds, as discussed in the review of Griffin et al. (2007), the classical method of GTG-banding, in general, is limited, due the poor distinction of chromosome bands compared to mammals and it is not possible to visualize bands in the microchromosomes. Therefore, the great insights about chromosome evolution in birds were made after the introduction of chromosome painting. Due to the extent of data already published about genomic comparison using chromosome painting in mammals and birds, the next section is separated in two topics in an attempt to describe the mammals and birds discoveries.

7.1. Mammals Karyotype Evolution

In recent years, the greatest revolution in comparative cytogenetic of mammals was obtained due to the increase of the number of species analyzed and the availability of probes for chromosome painting. Nowadays, most cases of Zoo-FISH among orders of mammals have been conducted referring to the human genome. For example, chromosome painting technique in mammal groups was applied onto genomes of about 150 species, being the human probes applied onto the chromosomes of ~70 mammal species (revised by Ferguson-Smith and Trifonov 2007, revised by Graphodatsky 2007, revised by Yang and Graphodatsky 2009). In the review performed by Ferguson-Smith and Trifonov (2007), it is related that most of these studies revealed mapping homologies among chromosome segments, thus allowing the identification of karyotype rearrangements among the main mammal lineages and their evolutionary interrelationship. In other words, Zoo-FISH using human probes in species of each mammal order helped to complement and extend the phylogenetic relationships based on gene sequencing, as well as the establishment of an ancestral karyotype for primates, carnivores, cetartiodactyls, perissodactyls and rodents, and therefore, the ancestral eutherian karyotype (AEK).

As already discussed, it is unfeasible to apply the human chromosome probes onto monotreme (Prototheria, Monotremata) chromosomes. However, there are some studies using comparative BAC mapping and chromosome painting using monotreme chromosome probes onto their own chromosomesin an attempt to study sex chromosome (Grützner et al. 2004, Rens et al. 2004, 2007, revised by Graves 2008, revised by Ferguson-Smith and Trifonov 2007). Other chromosome probes which were isolated by the flow-sorting of platypus genome (Rens et al. 2004, McMillan et al. 2007) may increase the extent of data about monotremes karyotype evolution. For example, Zoo-FISH experiments using platypus (monotreme) chromosomes as probes onto chromosomes of short-beaked echidna showed a conservation *in toto* of 10 autosomes between both species and besides, others platypus autosomes represent chromosomal segments in echidna (monotreme) (for more details see Rens et al. 2007). Although these results are interesting for monotreme karyotype evolution, the major discussion that includes this group are about monotremes and therian (marsupials and eutherians) sex chromosomes evolution, as it will be shown in Chapter VIII.

The Marsupialia includes opossum and kangaroo species and the chromosome painting applied in this clade revealed homology between the X chromosomes of marsupials and eutherian mammals. In addition, the autosome homologies data is still lacking for this group; however, the chromosome painting using chromosomes of five marsupial species as probes showed that the Australian marsupials as well as American Didelphidae have similar genomes (Rens et al. 2001).

The Zoo-FISH using chromosome probes of eutherians (in general, human chromosomes) onto monotreme or marsupial chromosomes is not as easy as aforementioned, thus the great karyotypic comparisons using this technique is restricted to Eutheria clade (including the Xenarthra, Afrotheria, Laurasiatheria and Euarchontoglires superorders). These Zoo-FISH findings have allowed the comparison among syntenic blocks (which involve a whole chromosome or chromosomal segments) in almost all eutherian orders, often using the *Homo sapiens sapiens* chromosomes (HSA) as reference (revised by Ferguson-Smith and Trifonov 2007). Therefore, it was concluded that some syntenic associations (syntenic blocks) are specific of one group (they are the cytogenetic signatures of a clade) and other are conserved among two or more, or even so throughout the eutherian clade. These analyses helped to construct the ancestral eutherian karyotype (AEK) (Chowdhary et al. 1998, Frönicke et al. 2003, Richard et al. 2003, Yang et al. 2003, Robinson et al. 2004, Svartman et al. 2004b, Wienberg 2004, Froenicke 2005, Froenicke et al. 2006, for more details see review by Ferguson-Smith and Trifonov 2007).

Due to the great amount of data related to eutherian mammals (~90 species in ~18 orders), which are far outside the scope of this chapter, not all papers for all discovery for all orders will be discussed here (the extent of information about this theme into this clade is worthy of its own chapter). In this way, the reader is advised to check for papers cited in this section as well as other papers available in the scientific literature (for example, the list of articles in the Cambridge Resource Centre for Comparative Genomics, http://www.vet.cam.ac.uk/genomics/ or in others web search tools). Other information, such as the location of each ancestral eutherian synteny into some eutherian karyotypes, references of the important papers in this area, karyotypes and GTG-banding for many eutherian species, and an extensive chromosome homology map, can be retrieved in the CHROMHOME

(CHROMosome HOmology Mapping and E-painting) http://www.chromhome.org/. Thus, the great aim of this section is to show some of the conclusions obtained when the chromosome painting is applied in karyotype evolution, more specifically in the eutherian karyotype evolution.

The power of comparative chromosome painting, such as the establishment of phylogenetic relationships of ancestral karyotypes and cytogenetic signatures of one group, as already mentioned, can be noticed in studies using the eutherian species. For example, the establishment of the syntenic block 1/19p as a shared synteny between Afrotheria (aardvark, elephants, golden mole, manatee and elephant-shrew) and Xenarthra (tree anteater) (revised by Ferguson-Smith and Trifonov 2007) could be present in the common ancestor for Afrotheria and Xenarthra (compose the South American armadillos, anteaters, three-toed sloths and two-toed sloths) (Yang et al. 2006). In the same way, the phylogenetic relationships were also shown between the Eulipotyphla and Pholidota orders, in which four syntenic associations are shared between Javan Pangolin (Pholidota order) and long-eared hedgehog (Eulipotyphla) (Yang et al. 2006, Ye et al. 2006). Likewise, in the Chiroptera order, Zoo-FISH and GTG-banding analysis corroborated the division of the families of this order in three lineages and showed a close relationship between two of these lineages, the Megachiroptera (Pteropodidae) and Rhinolophoidea (Microchiroptera) (for more details see Volleth et al. 2002, Ao et al. 2007).

Since the first Zoo-FISH performed by Wienberg et al. (1990) in one primate species, this technique has been applied in more than 50 species of primates under studies of comparative chromosome painting (revised by Stanyon et al. 2008). It is evident that the chromosome painting was very important to presume the putative ancestral karyotype in many primates (see Table 2) and in the establishment of phylogenetic relationships in this group (Muller et al. 1999, O'Brien and Stanyon 1999, Murphy et al. 2001, revised by Ferguson-Smith and Trifonov 2007, revised by Stanyon et al. 2008). As an example of karyotype evolution of primates, chromosome painting showed immense homologies between all human chromosomes and apes chromosomes, except for a centric fusion between two chromosomes, which formed the HSA 2 (revised by Ferguson-Smith and Trifonov 2007).

Interestingly, besides all results produced by Zoo-FISH and BAC mapping experiments, the comparative chromosome painting in hominoids karyotype evolution showed that the orangutan lineage diverged earlier than gorilla and chimpanzee lineage. After this, the gorilla lineage was the second to diverge and finally, the human line acquired three inversions and one chromosome fusion (the chromosome 2), which reduced the number of chromosome from 2n=48 to 2n=46 (see Table 2) (revised by Ferguson-Smith and Trifonov 2007).

The insights about the evolution of one group (as noticed above in apes and human cases) using chromosome painting can be also observed in Cetartiodactyla order, in which the hypothesis was corroborated that the Camelidae family was the first to diverge during the Cetartiodactyla evolution followed by divergence of Suina and finally of Cetacea/Hippopotamidae and Pecora groups (for more details see Kulemzina et al. 2009).

In addition, chromosome painting showed that the Afrotheria superorder (Frönicke et al. 2003, Yang et al. 2003, Robinson et al. 2004), Canidae order (Graphodatsky et al. 2002, Nie et al. 2002, Perelman et al. 2005, 2008, revised by Ferguson-Smith and Trifonov 2007), Perissodactyla order (include tapirs, rhinoceros, horses, onager, kulan, donkey, and zebras) (revised by Trifonov et al. 2008), Eulipothyphla order (include hedgehogs, shrews, solenodons and moles) (revised by Ye et al. 2006), Cetartiodactyla order (include cattle, sheep, deer, giraffes, pigs, camels, hippopotamus, dolphins and whales) (revised by Ferguson-Smith and Trifonov 2007) and Rodentia order (Froenicke et al. 2006) possess different rates of karyotype rearrangements into their evolutionary history (revised by Ferguson-Smith and Trifonov 2007).

Table 7.2. Predicted ancestral karyotypes for several primate groups based on the 23 human chromosomes

HSA	AHK	ACK	APLK	AAK	APK	AEK
1	1	1	1a, 1b, 1c	1	1	1
2	2a, 2b or 2p-q, 2q	2a, 2b	2a, 2b/16q	2p-q, 2q	2p-q, 2q	2p-q12, 2q
3	3	3	3a, 3b, 3c/21	3/21	3/21	3/21
4	4	4	4	4	4	8p/4
5	5	5	5/7a	5	5	5
6	6	6	6	6	6	6
7	7	7	7b, 5/7a	7a, 7b	7b, 7a/16p	7a/16p, 7b
8	8	8	8a, 8p/18	8	8	8p/4, 8q
9	9	9	9	9	9	9
10	10	10	10p, 10q/16	10p, 10q	10p, 10q	10p/12pq/22qt, 10q
11	11	11	11	11	11	11
12	12	12	12	12	12a/22a, 12b/22b	10p/12pq/22qt, 12qt/22q
13	13	13	13	13	13	13
14	14	14/15	14/15a	14/15	14/15	14/15
15	15	14/15	14/15a, 15b	14/15	14/15	14/15
16	16	16	10q/16p, 2p/16q	16p/16q	7a/16p, 16q	19q/16q, 7a/16p
17	17	17	17	17	17	17
18	18	18	8p/18	18	18	18
19	19	19	19	19	19p, 19q	19p, 19q/16q
20	20	20	20	20	20	20
21	21	21	3c/21	3/21	3/21	3/21
22	22	22	22	22	12a/22a, 12b/22b	12qt/22q, 10p/12pq/22qt
X	X	X	X	X	X	X
Y	Y	Y	Y	Y	Y	-
23	24	23	27	25	25	23

HSA, *Homo sapiens sapiens*; AHK, ancestral hominoid karyotype; ACK, ancestral catarrhine karyotype; APLK, ancestral platyrrhine karyotype; AAK, ancestral anthropoids karyotype; APK, ancestral karyotype of all primates; AEK, ancestral eutherian karyotype. On the bottom of the table are presented the diploid number of each karyotype. Data obtained from Ferguson-Smith and Trifonov (2007), and Stanyon et al. (2008)

In this way, the Canidae, Ursidae and Mephitidae families (all belonging to Carnivora order), and muroid superfamily (for example mice, hamster and rats) have higher rates of chromosome evolution (Nash et al. 1998, Yang et al. 1999, Fröenicke et al. 2006, revised by Ferguson-Smith and Trifonov 2007, Perelman et al. 2008), being that the latter group has the highest evolutionary rate among mammal species (revised by Graphodatsky 2007). In fact, comparative reciprocal painting between mice and rats revealed that the rate of divergence between both species is tenfold higher than the one observed between a human and a cat, and represents a good example of the high number of rearrangements in muroids (Stanyon et al. 1999).

Furthermore, the chromosome painting in eutherian species allowed obtaining the cytogenetic signatures for the superorder Afrotheria and Xenarthra (Yang et al. 2003, 2006, Robinson et al. 2004, Gilbert et al. 2006, Kellogg et al. 2007, Pardini et al. 2007) and on the other hand, for Euarchontoglires (composed of Primates, Scandentia, Dermoptera, Lagomorpha and Rodentia orders) and Laurasiatheria (composed of Eulipotyphla, Chiroptera, Pholidota, Carnivora, Perissodactyla and Cetartiodactyla orders) cytogenetic signatures were not identified (revised by Ferguson-Smith and Trifonov 2007) (for more details, see the discussion of syntenic association shared into Laurasiatheria superorder by Yang et al. 2006). In addition, the establishment of their respective ancestral karyotypes was obtained for some orders (revised by Ferguson-Smith and Trifonov 2007).

Comparative chromosome painting was applied in ~90 species in ~18 orders of eutherian mammals in an attempt to build an ancestral eutherian karyotype (Chowdhary et al. 1998, Frönicke et al. 2003, Richard et al. 2003, Yang et al. 2003, Robinson et al. 2004, Svartman et al. 2004b, Wienberg 2004, Froenicke 2005, Froenicke et al. 2006, for more details see review by Ferguson-Smith and Trifonov 2007). Throughout the years, this ancestral karyotype has been refined on the basis of cytogenetic studies as well as the analysis of bioinformatics (Froenicke et al. 2006, Robinson and Ruiz-Herrera 2008).

Based on syntenic associations that are shared by 18 mammal orders (it is thought that conserved syntenic association blocks could be present in ancestral karyotypes), it was possible to hypothesize the most likely ancestral eutherian karyotype, which had 2n=46 chromosomes (Froenicke 2005, Froenicke et al. 2006, revised by Ferguson-Smith and Trifonov 2007). After the completion of this ancestral eutherian karyotype, it was observed that the human karyotype has many features of this ancestral genome. For example, it was suggested that many human chromosome segments were not disrupted in AEK, which formed entire chromosomes in the ancestral (notably the entire human chromosomes 1, 5, 6, 9, 11, 13, 17, 18, 20 and X are conserved intact in this putative ancestral karyotype). Likewise, entire human chromosomes were connected or linked with part(s) of other human chromosome(s) that also formed entire chromosomes in AEK (Table 3) (revised by Ferguson-Smith and Trifonov 2007). Interestingly, the ancestral eutherian genomes models suggested by the bioinformatics and molecular cytogenetic have similar chromosome numbers, but surprisingly, the number of conserved segments and the number of syntenic associations is greatly different. Thus, the authors argued that the ancestral eutherian genome model suggested by cytogenetics may be a far more likely representation of this ancestral genome (Froenicke et al. 2006).

Table 7.3. Ancestral eutherian karyotype and its relationship to human chromosome syntenies

AEK	Human
1	1
2	* 8p/4 (4q/8p/4pq)
3	* 3/21 (3p/21)
4	5
5	* 14/15
6	6
7	* 10p/12pq/22qt
8	2q
9	7 (7b)
10	2p-q1.2
11	9
12	11
13	10q
14	13
15	8q
16	17
17	18
18	* 19q/16q
19	20
20	* 16p/7 (16p/7a)
21	* 12/22 (12qt/22q)
22	19p
X	X

AEK, Ancestral eutherian karyotype; (*) indicates the human syntenies conserved for most eutherians. Bioinformatics analysis revealed that some of these syntenies are also present in opossum and/or chicken, being probable synplesiomorphies for Eutheria (see Robinson and Ruiz-Herrera 2008). The syntenies into parenthesis are refinements of the syntenic associations. Data obtained from Froenicke et al. (2006), Yang et al. (2006), and Ferguson-Smith and Trifonov (2007). For more details see Robinson and Ruiz-Herrera (2008), among others

Interestingly, the analysis of whole-genome sequence assemblies between human and opossum (*Monodelphis domestica*), and between human and *Gallus gallus* (opossum and chicken are used as outgroups in this case), determined the regions of synteny among them. The analysis of these homologous syntenic blocks revealed the human chromosome 1, 5, 6, 9, 11, 17, 20, and the X (all intact chromosomes in AEK) are disrupted in opossum and chicken (Robinson and Ruiz-Herrera 2008). The meaning of this data is that these characteristics are monophyletic for eutherian clade.

Moreover, these sequence comparisons also concluded that the human chromosomes 13 and 18 (both also intact chromosomes in AEK) are also intact in chicken and opossum karyotypes, respectively, and it means that these karyotype features are symplesiomorphic (shared ancestral). Moreover, it was also concluded that the human segments 8q, 10q and 19p, which are also present in AEK (Table 3), are present as a single syntenic block in opossum. In the same way, the 8q and 10q are also present as a single block in chicken, and thus, these segments (8q, 10q and 19p) represent symplesiomorphic characteristics for Eutheria, but they are not monophylic characters for eutherian in relation to the outgroups (because of the sharing among these species) (for more details see Robinson and Ruiz-Herrera 2008).

It is remarkable that chromosome painting was a very important cytogenetic tool for the establishment of karyotype relationships or corroborating previous phylogenetics hypothesis in these eutherian orders as well as others not cited here. Moreover, chromosome painting refuted or corroborated many putative genomic blocks based in GTG-banding shared among mammalian groups. In other words, it means that chromosome painting has more resolution than classical cytogenetic methods. On the other hand, despite the great contribution of the comparative chromosome painting about evolution of mammal karyotypes, caution is needed in the analysis and comparisons of the syntenic associations because, as revised by Ferguson-Smith and Trifonov (2007), a syntenic association may be shared by two or more groups due to convergence, that may occur by centromeric fusion and/or fission, for example.

7.2. AVIAN KARYOTYPE EVOLUTION

Although most studies about comparative chromosome painting are related to mammals, there are also many studies applied to avian groups. The birds are separated in two large groups known as Paleognathae (ratites) and Neognathae (carinates), and in general, the number of chromosomes range from 2n=74-86 and 2n=66-74, around 63% and 24% of species, respectively (in general 2n~80 chromosomes). The karyotype characteristic in birds includes macrochromosomes, sex chromosomes and several microchromosomes. The general karyotype pattern of birds was maintained relatively constant along the diversification of the groups (revised by Griffin et al. 2007).

Shortly after the development of chromosome-specific probe for each *Gallus gallus domesticus* macrochromosome, including nine autosomes and the Z chromosome (Griffin et al. 1999), Shetty et al. (1999) conducted the first Zoo-FISH in birds. In this remarkable study, the 10 chromosomes of chicken (Neognathae) were used as probes onto the emu *Dromaius novaehollandis* (Paleognathae) chromosomes, both distantly related. In general, the results showed that each chromosome probe of chicken hybridized in the same chromosome in emu and these results revealed a great homology between the two species, corroborating the hypotheses of karyotype conservation suggested by classical cytogenetic methods, which generally includes the macrochromosomes and the sex chromosome.

Some years after the first Zoo-FISH in birds, Guttenbach et al. (2003) conducted Zoo-FISH using chicken chromosome probes over the chromosomes of Paleognathae *Rhea americana* and the results also revealed an overall karyotypic conservation in avian groups. Thus, it was reported that the divergence of Paleognathae and Neognathae (the first great divergence in avian group) was not defined by remarkable and specific karyotype change in the macrochromosomes (except by arising of heteromorphic sex chromosomes in Neognathae). In fact, the thought that the bird karyotypes are stable, apparently higher than mammals (Burt et al. 1999), has been supported in recent studies (Shibusawa et al. 2004 a, b, revised by Griffin et al. 2007, Nanda et al. 2008). During birds' evolution, the karyotype conservation and the absence of intense chromosomal changes were also maintained after the divergence of Galliformes and Anseriformes (both Neognathae birds). Likewise, another important evolutionary event in birds was the divergence between 'higher land' and 'higher water' birds and this process was not followed by major karyotype changes, except for a series of microchromosomal fusions which tended to reduce the diploid number in some groups (revised by Griffin et al. 2007). On the other hand, some studies with Zoo-FISH using chicken probes has showed an excessive chromosome rearrangement in Falconiformes and Psittaciformes birds (for more details see Nanda et al. 2006, 2007, 2008).

Although the first Zoo-FISH was carried out in 1999 (Shetty et al. 1999), the first draft of the ancestral avian karyotype was described many years ago by Stock and Bunch (1982) in examining banding patterns. Based in Zoo-FISH analysis, Shibusawa et al. (2004a) describes the first ancestral karyotype for some avian groups (Galliformes) and predicted a conserved karyotype for the avian lineages. After this, the search for an ancestral karyotype of avian group continued to be refined and involved many studies of Zoo-FISH and even sequence comparison between chicken and soft-shelled turtles, and between chicken and humans. Nowadays, it is thought that the ancestral karyotype of the avian group is composed of 10 macrochromosomes and a Z chromosome (excluding the microchromosomes), being the *G. gallus domesticus* karyotype, excluding the microchromosomes, very similar to this ancestral karyotype. In fact, the only difference between both karyotypes is that the chicken chromosome 4 is a product of the fusion between 4 (4q of *G. gallus domesticus*) and 10 (4p *G. gallus domesticus*) ancestral avian chromosomes. In general, the studies reveal that the chicken chromosomes 1, 2, 3, 4q, 5, 6, 7, 8, 9, 4p and Z represents the ancestral chromosomes 1-10 + Z for all birds. Interestingly, based on sequence analysis, the ancestral avian chromosome 4 is also present in mammals and turtles, representing an ancient chromosome. Later in the evolutionary history of vertebrates, the chromosomes 1, 2, 3, 5 and Z arise and the sharing of these chromosomes between turtles and birds suggests that the emergence occurred before the divergence of both animal groups (for more details see the review by Griffin et al. 2007).

Finally, in all cases aforementioned for birds karyotype evolution (for more details see the review by Griffin et al. 2007), the Zoo-FISH using chromosomal probes of *G. gallus domesticus* refined and corroborated the previous hypothesis to avian karyotype evolution proposed by classical cytogenetic methods.

All data available about chromosome painting in mammals and birds reveals that, in general, the mammals have more karyotypic changes than birds (Wienberg 2004), or in other words, the birds have higher karyotype stability than mammals (Burt et al. 1999, Shibusawa et al. 2004 a, b, reviewed by Griffin et al. 2007, Nanda et al. 2008). Good examples of these affirmation may be noticed in the Chinese Muntjac (2n=46), Indian Muntjac (2n=6 and 7 in males and females, respectively) (Yang et al. 1997b), gibbons, Muridae rodents, caniforms (Wienberg 2004), equides (Trifonov et al. 2008), among others, which generally had many karyotype rearrangements during their evolutionary history. Some of these conclusions were achieved a long time before chromosome painting advents, but there are no doubts that painting methodology refined and confirmed these conclusions.

7.3. The Future of Chromosome Painting

Chromosome painting is an excellent approach for the elucidation of origin, evolution and relationship between different chromosomes and genomes. The most notable contributions of chromosome painting seem to be the establishment of an ancestral karyotype, phylogenetic relationships and the probable origin of some specific chromosomes. Cross-species chromosome painting is very resolute to identify homologies among syntenic blocks, thus allowing the construction of karyotype phylogenies between species, families and orders, in addition to the establishment of their ancestral karyotypes, as observed in mammals and birds. Moreover, the data obtained by chromosome painting can be integrated to other data, such as the analysis of heterochromatin patterns, BAC-FISH mapping, FISH with other kinds of probes as well as whole- genome sequences comparisons.

The application of chromosome painting for comparative chromosome studies in diverse animal groups is expected, despite the difficulties for probe obtaining in some specific groups. Furthermore, the chromosome comparisons can be predicted comparing available nucleotide sequences of whole genomes. In this way, bioinformatics seems to be an important tool to be integrated to classical painting and applied in cytogenetics, generating data on the chromosomal level based exclusively on nucleotide sequence data.

Chapter 8

THE "ODD" CHROMOSOMES

8.1. SEX CHROMOSOMES

The reports about sex determination knowledge are much more ancient than the emergence of cell theory and along the years it became a subject of discussion among philosophers, thinkers and scientists. Nowadays, it is known that the sex determination is very important for reproduction, being substantial for evolution and genome diversity. In general, it is thought that the sex determination in animals may be consequence of at least three mechanisms: (i) Environmental Sex Determination (ESD), in which the environment influences sex determination of embryos at a specific time of development. The effect of temperature is the most understood parameter and in this case, it acts on the stage of development named thermosensitive period (TSP), being the trigger of the sex determination and differentiation. These mechanisms have been noticed in several fish and reptiles; (ii) Genetic Sex Determination system (GSD), in which there is the presence of distinct sex chromosomes which probably harbor a major sex determining gene (e.g. Sry of therian or Dmy of medaka *Oryzias latipes*) (Devlin and Nagahama 2002, Matsuda 2005, revised by Manolakou et al. 2006, Wallis et al. 2008, revised by DiNapoli and Capel 2008, revised by Graves and Peichel 2010); (iii) Genetic action, in which there is at least one specific gene that is considered the main regulator of sexual phenotype development pathway (noticed for example in insects, worms and amphibians) (revised by Manolakou et al. 2006).

Although there are many systems of sex determination, the pathways that lead any bipotential gonads towards maleness or femaleness are better understood in therian mammals than in other animals. In this group, the Sry gene, which is localized in the Y chromosome, codify one testis determinant factor (Sinclair et al. 1990, Koopman et al. 1990) and its expression conduct to male traits development by a complex network that includes the regulation of many genes (for more details see DiNapoli and Capel 2008, revised by Graves and Peichel 2010). Interestingly, the sex-determining gene, showing a similar function to Sry, has not been found in non-therian species except in the medaka species, which will be discussed in this topic.

Taken together, the knowledge about sex determination in mammals may create an idea of "rule of sex determination", in which every animal must have a evident sex chromosome harboring one gene that triggers the male or female development. However, the studies in a range of animals have shown that this "rule" is, in fact, the exception. In other words, the variability of sex determination mechanisms is wider, diverse and more divergent across the non-therian species than existent in therians animals. For example, in teleosts it is observed the influence of autossomal genes, hormone secretion, epigenetic factors, environmental factors, behavior influence, and several sex chromosomes systems (XX/XY, ZZ/ZW, X_1X_2/X_1X_2Y, XX/XY_1Y_2, ZZ/ZW_1W_2). Moreover, some species have several kinds of hermaphroditism besides sex reversal in several stages of their life (revised by Devlin and Nagahama 2002, revised by Manolakou et al. 2006, revised by Mank et al. 2006). In addition, the insects, another group with a richness of species, posses XX/XY, ZZ/ZW, haploid/diploid system (Hymenoptera and Isoptera) and X:autosome ratio system (Dipterans and Nematodes) (revised by Manolakou et al. 2006, revised by Uller et al. 2007, revised by Janousek and Mrackova 2010, revised by Verhulst et al. 2010).

It is broadly reported in the literature the presence of sex chromosomes in a range of animals. In some organism, such as therian mammals, the role of these chromosomes in sex determination is well known. However, in general, there is a wide gap between the description of one sex chromosome under the light of a microscopy and the identification of its evident role in the sex phenotype development. Anyway, the sex chromosomes have attracted the interest of many researchers due to their potential role or due to their traits and in this way, the study of sex chromosomes may increase the knowledge about their evolution as well as their influence in the mechanisms of sex determination and differentiation. Many articles have been published along the years about sex chromosomes mainly with respect to their evolution and description.

The accumulated knowledge makes it possible to create a theory about the origin of sex chromosomes that has been applied to explain the origin of these chromosomes across the kingdoms (fungi, plants and animals) (Fraser and Heitman 2005). In general, it is known that the trigger events of sex chromosome evolution could occur in any part of any chromosome, however, there are evidences that they evolve from a specific pair of autosomes carrying sex-determining gene(s). In this way, the steps of this process are the suppression or reduction of recombination, chromatin differentiation, and degeneration of one homologue of the pair (revised by Ming and More 2007, revised by Janousek and Mrackova 2010). Finally, all process aforementioned generally conduct to the arising of two specific regions in these sex chromosomes, known as: pseudo-autosomal regions (PARs), a region of recombination between both homologous (XY or ZW chromosomes); and a male-specific regions of the Y chromosome (MSY) or female-specific regions of the W chromosome (FSW), accordingly to sex chromosome system (for more details see Ming and Moore 2007, and references therein).

The suppression/reduction of recombination among regions of the sex chromosomes is one of the first and main steps to sex chromosome evolution. In the review performed by Charlesworth et al. (2005), it was argued that the main factors that conduct to the reduction/suppression of recombination are the presence of sex determining genes without recombination among the homologous of one pair and the evolution of alleles that are beneficial to one sex. In summary, to trigger the evolution of sex chromosomes, it is necessary to have differences in DNA sequences among some loci located in homologous of the "pre-sex chromosomes" (Figure 8.1).

Charlesworth et al. (2005) discussed that in the "proto-sex chromosomes" (the authors use the terms "proto-Y" and "proto-X", however, the mechanism is similar for XY and ZW systems) the differentiation occurs on genes that possess interaction with the primary sex determining genes, allowing the arising of new alleles beneficial to one sex (Figure 8.1, 1st stage). Once such genes have accumulated high differences, the suppression of recombination along a wider genomic region on these "proto-Y/W" occurs. Moreover, other mechanisms, such as chromosome rearrangements or insertion of transposable elements (TEs), may influence the recombination rate of the sex chromosomes (Figure 8.1, 2st stage). Furthermore, genetic changes such as the arising of new alleles biased to one sex may also occur, besides the consequently increase of the non-recombining area (Figure 8.1, 3st stage) (for more details see Charlesworth et al. 2005). In other articles (see Ming and Moore 2007), it was argued that the suppression/reduction of recombination is started by a chromosomal rearrangement (such as insertions of any specific region) near the region harboring any sex-determining gene(s), producing a hemizygous state. Soon after, there is the expansion of the differential region (e.g. MSY or FSW) resulting in the increase to reduction/suppression of recombination around the sex determining region (revised by Ming and Moore 2007, revised by Janousek and Mrackova 2010).

Furthermore, along the early steps of suppression/reduction of recombination, the homologous chromosomes recognized as sex chromosomes are morphologically indistinguishable and the differentiation of one homologue of the sex chromosome pair (Y or W) is trigged by this mechanism. Soon after, the accumulation of TEs occur (Charlesworth et al. 2005, Ming and Moore 2007). Interestingly, it is predicted that all sex chromosomes are, in fact, heteromorphic chromosomes, at least at the molecular level (revised by Ming and Moore 2007). In other words, cytologically homomorphic incipient sex chromosomes (known as chromosomes in early stages of evolution, commonly noticed in fishes, amphibians and lizards) may posses a sex-specific region in one homologous of the pair.

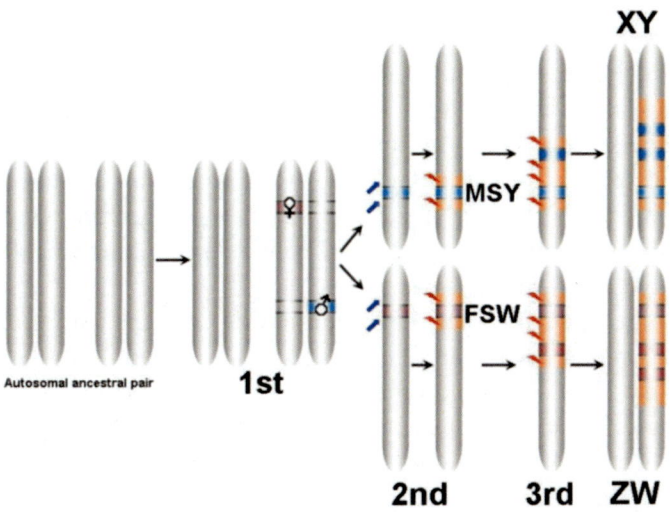

Figure 8.1. Scheme of heteromorphic sex chromosome evolution from an autosomal pair. Genes/alleles that are beneficial to male (blue bars), female (red bars), original alleles (empty bars). The orange bars indicate chromosomal regions with reduction/suppression of recombination. The blue arrows indicate the regions that will be subject to reduction/suppression of recombination. Red signals represent the insertion of TEs.

Regarding the emergence of the visible heteromorphic sex chromosomes, genetic degeneration necessarily occurs (reduction of ability of selection to fix favorable mutation and to prevent the fixation of deleterious mutation) or extension of the chromatin of one homologous of the sex chromosome (Y or W). In this process, TEs may be responsible to conduct the degeneration among most genes present in Y or W chromosomes (Fraser and Heitman 2005, revised by Charlesworth et al. 2005, for more details see Ming and More 2007, revised by Janousek and Mrackova 2010). This process is better understood in human Y chromosome and this knowledge leads us to think that it is a trait of a Y chromosome (or W chromosome). However, this is not the scenario for a range of animals and plants. For example, it is predicted that Y or W chromosomes could disappear without the steps of reduction in size, and thus emerging a X:autosome or Z:autosome system (revised by Ming and Moore 2007).

The use of molecular cytogenetics has produced a wide quantity of information concerning sex chromosomes in animals, mainly vertebrates such as mammals and birds, and in the lesser extent, in fish, amphibians and reptiles. The Zoo-FISH using HSA X chromosome as a probe, as well as other cytogenetic methods applied throughout Eutheria clade, revealed that this chromosome is conserved among all placental mammals and probably was present in the AEK (Table 3, Chapter VII) (revised Ferguson-Smith and Trifonov 2007 see other papers cited in the beginning of Chapter VII). An interesting fact was observed when the HSA X was used as a probe onto a two-toed sloth (Pilosa, Xenarthra), showing signals in the entire X and in part of Y chromosomes of this species, indicating the existence of some homology between these two chromosomes. It was surprising because these patterns of hybridizations using human probes had never been reported in mammals, with the exception of the cross-hybridization to the PARs between closely related species (Yang et al. 2006).

Another interesting characteristic in relation to the mammal sex chromosomes was observed in a Monotreme species named duck-billed platypus (*Ornithorhynchus anatinus*). The chromosome painting of individual chromosomes of platypus revealed that they have ten sex chromosomes (X_1-X_5/Y_1-Y_5) and, interestingly, these chromosomes are organized in a chain configuration at male meiosis (Rens et al. 2004, revised by Grützner et al. 2004). These sex chromosome traits, as well as the enigmatic platypus genome characteristics (Brown 2008), made the sex chromosome systems of Monotremes interesting for the comparative chromosome painting analysis.

The comparative gene mapping analysis showed that the largest X chromosome of platypus (X_1) has homology with part of eutherian and marsupial X chromosomes, and the X_5 has homology to the birds Z chromosome. The whole chromosome painting of platypus sex chromosomes showed that the sex chromosome chain evolved from an original sex chromosome pair with homology to the avian ZW system and during the early evolution of mammals, four autosome pairs were translocated for this system (Grützner et al. 2004, revised by Graves 2008). Many years before this conclusion, Ohno (1967) proposed that both **mammals' and birds' sex chromosome systems have evolved from the same autosomal pair. Ohno's proposition** was refuted after the demonstration that the chicken Z chromosome is homologous to the human chromosomes 9 and 5, and that the X chromosome of therian is homologous to chicken chromosome 4p and part of chromosome 1 (Graves and Shetty 2000, Nanda et al. 2000). In other words, the sex chromosomes of both groups were derived from different pairs of autosomes of the common ancestor (Nanda et al. 1999, 2000). Nevertheless, the recent insights about sex chromosome evolution in the platypus evoked Ohno's theory (1967), due to the observation of the homologies between sex chromosomes of platyplus with the therian XY and birds ZW system (Grützner et al. 2004).

This 'link' was questioned when two new papers using comparative gene mapping about this theme were published (Rens et al. 2007, Veyrunes et al. 2008). These articles showed that platypus X_1 and X_3 have homology tosome chicken autosomes and Z chromosomes and, some chicken Z genes are present on X_3 and X_5 chromosomes of platypus and also on X_3 and X_4 chromosomes of echidna (another monotreme). Either way, X_1 of platypus and eutherian X chromosomes do not share homologies, and consequently it was proposed that the X_1 of platypus is more related to the avian chromosomes than to therian sex chromosomes (Rens et al. 2007, Veyrune et al. 2008). Taken together, the authors argued that these results may indicate that the monotreme sex chromosome system could be linked to the avian sex chromosome system by an ancestral sauropsid sex chromosome system (or an ancestral bird-like ZW sex chromosome) and that, in fact, they are not a link between bird and mammal sex chromosome systems, as proposed before by Grützner et al. (2004) (Rens et al. 2007, Veyrunes et al. 2008). Consequently, it was proposed that the therian sex chromosome system evolved after the prototherian and therian divergence in which the amniote ancestor may have had a bird-like ZZ/ZW system and thereafter, on therian lineage the XY system has evolved (Rens et al. 2007, Veyrunes et al. 2008). In summary, it was concluded that the monotremes and eutherian sex chromosomes had independent origins, such as currently observed for sex chromosome evolution in teleost fish (revised by Mank et al. 2006).

Recently, the sex chromosomes of the short-beaked echidna (*Tachyglossus aculeatus*) and platypus were compared by Zoo-FISH experiments. At first, it was concluded that the echidna has 10 sex chromosomes in females (ten Xs) and 9 in males (five Xs and four Ys), and, similarly to the sex chromosomes of platypus, the sex chromosomes of echidna are organized in chain configuration at meiosis. Moreover, the results revealed that some chromosomes of platypus (X_1-X_3/Y_1-Y_3) are conserved *in toto* on echidna karyotype and that some other sex chromosomes are homologous for some sex chromosomes and autosomes to the echidna and *vice-versa* (for more details see Rens et al. 2007). In addition, it was shown that the 'Y_5 chromosome' present in platypus but absent in echidna (it was proposed that the Y had been lost in this species) was incorporated into Y_3 in the latter species. Finally, other experiments (including some reciprocal chromosomal painting) showed that, although both platypus and echidna sex chromosomes are organized in meiotic chains, all data indicates differences in the constitution and order of the chromosomes between both chains, and they have probably continued to evolve after the divergence between echidna and platypus (for more details, see Rens et al. 2007).

In relation to the sex chromosome in birds (all birds have ZZ/ZW system), the first paper about comparative chromosome painting in the avian clade sheds light on their origins. It was observed in a karyotypic conservation between Paloegnathes and Neognathes species models (emu and chicken, respectively). Interestingly, the *G. gallus domesticus* Z chromosome (GGAZ) probe hybridized both Z and W chromosomes of emu (the sex chromosome is a homomorphic pair in this species), showing a conservation of the Z chromosomes between two species and maintenance of an ancient conservation between Z and W sequences in birds. Consequently, the results of these studies corroborated the theory of autosomal origin for the sex chromosomes in birds group (Shetty et al. 1999).

Since this first article (Shetty et al. 1999), many other studies have been published related to molecular cytogenetics in avian sex chromosomes. These papers used comparative mapping with GGAZ-orthologous genes, comparative chromosome painting with GGAZ-specific DNA libraries, Zoo-FISH with GGAZ as probes and other probes which were hybridized in some Paleognathes and Neognathes species. Except for some special particularities of each study, the results generally supported and corroborated the evidence of the conservation of Z chromosome and an extensive homology between Z and W chromosomes, although the W chromosome is not more conserved than Z (Shetty et al. 1999, Shibusawa et al. 2004a, Nishida-Umehara et al. 2007, revised by Griffin et al. 2007, revised by Stiglec et al. 2007b, for more details see Nanda et al. 2008, among others).

Thus, the conservation of sex chromosomes into avian group may reflect the fact that birds have more of an overall karyotype stability than mammals (Nanda et al. 2008). It is also possible that the existence of some homology between Z and W chromosomes in birds and the absence of a general homology between X and Y chromosome (except for the sloth case) may reflect a higher karyotype stability of birds compared to mammals.

Some conclusions about the sex chromosome evolution of the Z and X chromosomes, in birds and eutherians, respectively, are similar. The sex chromosomes of homogametic sex (Z and X) shows little variability into some birds and mammals. On the other hand, several studies, as cited in the review by Griffin et al. (2007), did not find orthologues of the chicken W chromosome in other birds, and this statement has been also reported for the Y chromosome, indicating that the heterogametic sex chromosomes could show higher evolutionary rates.

In relation to the reptiles (excluding birds) sex development, they posses a great variability of sex determination ways, being that the ESD mediated temperature was the better understood mechanism. Moreover, the sex chromosomes are also reported in some groups. For example, snakes posses a conserved ZW sex chromosome system (revised by Manolakou et al. 2006) and lizards have XY, XXY, ZW and ZZW systems or a cryptic mechanism, besides the influence of temperature according to species (ESD and GSD may coexist in some groups) (for more details see Ezaz et al. 2009a). On the other hand, other reptiles, such as crocodiles, some terrestrial and sea turtles, do not posses any discernible sex chromosome. In the latter group, the temperature is a very important factor in sex determination (revised by Manolakou et al. 2006, revised by Graves 2008).

There are some karyotypes traits of snakes, lizards and some turtles similar to the avian karyotype (such as the presence of macro- and microchromosomes). These traits conducted the emergence of the hypothesis of a similar origin to the different ZW sex chromosomes of these different groups. In fact, in the case of *Gekko hokouensis* species, there is a homology between its ZW system and Z chromosome of chicken and this homology is reported as retention of ancient ZW sex chromosomes (Kawai et al. 2009). On the other hand, several approaches elucidated that generally, this is not the case. Thus, sex chromosome systems observed into these different groups have independent origins (including the different sex chromosomes system into a group, as noticed in lizards (see Ezaz et al. 2009a)) (Ezaz et al. 2009a, b, O'Meally et al. *in press*) and similar to that broadly reported in teleostei groups.

For example, it was shown that the chicken Z chromosome is homologous to part of chromosome 2 in snakes and that the Z of snakes is homologous to the short arm of chicken chromosome 2 and 27 (Matsuda et al. 2005; Matsubara et al. 2006; Kawai et al. 2007). Interestingly, a recent article (O'Meally et al. *in press*) showed that the W chromosomes of derived snakes share some repetitive sequence with the sex chromosomes of chicken, which could create the idea of common origins for both chromosomes. Nevertheless, the authors argued that these sequences accumulated independently on these chromosomes (convergent degeneration). In addition, it was noticed that some genes mapped in the Z chromosome of some snake species are also present in the autosome 6 of some lizard species (Ezaz et al. 2009b, Srikulnath et al. 2009). Moreover, the cross-species painting using the GGAZ probe showed a high homology to the chromosome 5 of female turtle *Chelodina longicollis* (Graves and Shetty 2001). Similar results were obtained when this probe was applied in the turtle *Trachemys scripta* and in the crocodile *Crocodylus niloticus*, showing homologies with the chromosome pair 6 of both species (they also do not possess a known sex chromosome) (revised by Ferguson-Smith 2007). In spite of independent origins of sex chromosomes in reptiles, the discussion about the precise origin of sex chromosomes was addressed in some articles. For example, the Dmrt1 gene was mapped in the chromosome 6 of species *T. scripta* and *C. niloticus* and in the Z chromosome of birds and taken together with the results of cross-species chromosome painting (already mentioned above), it was predicted that this chromosome (chromosome 6) represents the ancestral pair of chromosomes from which the sex chromosomes of birds have evolved (revised in Ferguson-Smith 2007).

In amphibians, the presence of XY and ZW sex chromosomes were noticed, commonly being homomorphic chromosomes. The influence of temperature on sex determination in several amphibian species is also reported (revised by Graves and Peichel 2010). Moreover, it is predicted that the ZW system is the original heterogametic sex for the amphibian group and the ZW and XY were interchangeable, suggesting that the sex chromosomes arose several times. In addition, the ancestral Z of the amphibian does not share ancestrality with the Z of birds or snakes (revised by Nakamura 2009). Moreover, the location of orthologs from GGAZ, GGA2 and GGA27 in *Xenopus* did not show a high conservation among both species (for more details see O'Meally et al. *in press*).

A curious case of sex chromosome may be observed in the pond frog *Rana rugosa* which has XY or ZW system, depending on the population (Nishioka and Hanada 1994, Nishioka et al. 1994, revised in Nakamura 2009). Moreover, the Y and Z, as well as X and W chromosomes, share common traits. Thus, several data, including cytogenetic mapping and sequence analysis, showed that the Z shares its origin with Y and the X with the W. It is speculated that the change in the sex chromosome system occurred by a pericentromeric inversion (revised by Nakamura 2009, revised by Janousek and Mrackova 2010).

In addition, in the amphibian *Xenopus laevis,* the presence of a sex-determining gene was evidenced (as described for medaka and therians), named W-linked DM-domain gene (DM-W). It is a similar copy of Dmrt1 gene localized in the W chromosome of this species and it is predicted to induce the ovary formation by antagonizing the Dmrt1 function and repressing the testis development. The DM-W must be a recent gene arisen in the amphibian sex determining pathways, because no female-specific copy of this gene was found in the *X. tropicalis* (a close related species of *X. laevis*), for example (Yoshimoto et al. 2008, revised by Nakamura 2009).

Regarding sex chromosomes in fish, more than 10% of species cytogenetically analyzed have identifiable sex chromosomes (including cartilaginous and bonny fish) (Devlin and Nagahama 2002). Fish groups have several sex chromosome systems but just in the medaka species a major sex determining gene was identifiable, which was the first sex-determining gene found in non-therian vertebrates. This gene is named Dmy (DM-domain gene on the Y chromosome) and conducts to male sex determination in this species (Matsuda et al. 2002, Matsuda 2005).

There are a great amount of published articles available in the literature in which the sex chromosomes of fish is the focus of discussion. The researches include classical cytogenetic analysis (Giemsa staining, C-banding and sinaptonemal complex analysis), cytogenetic mapping of DNA sequences, linkage mapping analysis, among other genomic approaches. In most of the studies, the classical traits of the sex chromosomes have been described (reduction of recombination, presence of heterochromatin and enrichment of repetitive elements), and an absence of a master sex-determining gene. Moreover, most of these studies are in agreement with many hypotheses about the sex chromosome evolution, such as an autosomal origin, lack of homologies between Z and W or X and Y, conservation of Z as well as X chromosomes throughout some specific taxa and multiple origins of each system.

For example, in the South American fish of the genus *Eignmannia*, the species *E. virescens* has XX/XY sex chromosome system (Almeida-Toledo and Foresti 2001), while another species named *E.* sp.2 has a X_1X_2Y sex chromosome system (Almeida-Toledo et al. 1988). The chromosome painting using the X chromosome of *E. virescens* and Y chromosome of *E.* sp.2 (E2Y) as probe onto their own genomes, showed that, in general, the X chromosome of *E. virescens* is homologous to the Y chromosome and part of chromosome 8. In the same way, the Y chromosome of *E.* sp.2 has homology with both X chromosome of its own genome. The Zoo-FISH, using the E2Y as a probe onto *E. virescens* chromosomes, revealed that the E2Y chromosome has high similarities with autosomal pairs of the *E. virescens*. The last results indicate a lack of homologies between both sex chromosome systems and reveal an independent evolution of these systems in these populations (Henning et al. 2008). In addition, it is probable that the X and Y chromosomes of the *E. virescens* originated from an autosomal pair.

Interestingly, in *Triportheus*, another South American fish that harbors a ZZ/ZW sex chromosome system, the cross-species chromosome painting performed using the Z chromosome of *T. nematurus* as a probe onto the chromosomes of several species of the genus, as well as into the chromosomes of some species of some genera putatively related to *Triportheus*, corroborated previous hypothesis of Z conservation into *Triportheus* genus beside an independent origin of sex chromosomes in this genus in relation to other genera. The results corroborate the synapomorphism of the sex chromosome systems for this genus (for more details see Diniz et al. 2008).

An interesting case of sex chromosome in fish was reported in the Nile tilapia fish, *Oreochromis niloticus*. It is believed that this species has a XY sex chromosome system (largest chromosome pair), proposed by synaptonemal complex analysis (Foresti et al. 1993, Carrasco et al. 1999). In this way, Harvey et al. (2002) created chromosome probes of X and Y chromosomes of the Nile tilapia and showed that these probes are specific to the largest chromosome pair, and that there are differences in sequence content between both chromosomes. Additional studies (Carrasco et al. 1999, Griffin et al. 2002, Harvey et al. 2003b, Ferreira and Martins 2008) pointed out that this chromosome pair has characteristics of a real sex chromosome, but the linkage mapping analysis, using BAC probes with sex determining markers, revealed that the sex determination locus related to the XY system is located in another chromosome pair (a small chromosome pair). In addition, the largest chromosome pair has a ZW locus for sex determination (Lee et al. 2003, Lee and Kocher 2007, Cnaani et al. 2008). Therefore, Cnaani et al. (2008) proposed that the ancestral group of

Oreochromis spp. had a ZZ/ZW system and in the *O. niloticus* lineage, the XY locus system took the control of the sex determination pathways.

Finally, it is notorious that the sex chromosomes in the therians and birds clades (except ratites birds) have a monophyletic origin and in these cases, a conservation of X and Z may be noticed, as well as a great divergence of Y and W, among the different species. A similar scenario may be also noticed in ZW of snakes. Intriguingly in other vertebrates, such as lizards, amphibians and fish, in general, there is not even a conservation of the sex chromosomes system, even among closely related species. Thus, in these groups, it is acceptable to conclude that the sex chromosomes have independent origins. Interestingly, although the sex chromosomes have multiple origins, the genomic forces that act in their evolution seems to be similar in the different species.

The studies about the origin and evolution of sex chromosomes are not only restricted to the vertebrate taxa, although they are the main focus of this chapter. In fact, there are many articles about this theme in invertebrate groups, although most of these studies were based on traditional cytogenetic methods, such as conventional analysis and C-banding. There are few works that use other cytogenetic approaches in the investigation of invertebrate sex chromosomes, such as chromosome painting and the use of other molecular cytogenetic tools, in an attempt to increase the knowledge about chromosome evolution. For example, in the genus *Dysdercus* (Heteroptera), the origin of the neo-XY sex system was investigated using 18S rDNA probe, GISH and Zoo-FISH of X chromosome obtained from *D. albofasciatus* (neo-XY) in comparison with the karyotypes of *D. chaquensis* and *D. ruficolis* with XO sex system (Bressa et al. 2009). The chromosome paints of an X chromosome together with rDNA mapping corroborated the previous idea proposed by Bressa et al. (1999) concerning the fusion of the ancestral X chromosome in *D. albofasciatus* with ancient XO sex-system generating the neo-XY bivalent. Moreover, this work (Bressa et al. 2009) added some refined information about the insertion of an X chromosome in the autosomal pair harboring rDNA cluster, followed by the inversion and transposition of rDNA sites from the neo-Y to the neo-X.

A good example of another kind of chromosome painting to study the sex chromosomes was performed by Willhoeft et al. (1998) using partial chromosome painting (PCP) of the Y chromosome (five probes, medY1-medY-5) and X chromosome (medX1) of the fruit fly *Ceratitis capitata*. The results, using the probes against the *C. capitata* chromosomes showed that the sex chromosomes in this species are composed of three regions in which: the region II (part of Yq, part of Yp and Xq) is the oldest thereby being the most degenerated region; the region I (centromeric region of X and Y, Xp, part of Yp and part of Yq) shows high homologies between the X and Y chromosomes; the region III (part of Yq and part of Xq) was the latter region to be inserted into the sex chromosomes. Moreover, Zoo-FISH using medY1, medY2 and medX1 probes onto chromosomes of *C. rosa* and *Trirhithrum coffeae* demonstrated that *C. capitata* is more related to *C. rosa* than to *T. coffeae*, and that at least some sequences of the Y chromosome of *C. capitata* were present in its own genome before the divergence between *C. capitata* and *C. rosa* (for more details see Willhoeft et al. 1998).

Another example is related to the ZW sex chromosomes in the clodling moth *Cydia pomonella*. The W chromosome of this species was used as a probe onto its own chromosomes, thus showing that this probe was a specific W chromosome probe. The absence of signals in the Z chromosome supports the evidence of a high level of molecular differentiation between Z and W chromosomes (for more details see Fuková et al. 2007).

In *Ephestia kuehniella*, *Cadra cautella*, *Plodia interpunctella* and *Galleria mellonella* species (Lepdoptera, Pyralidae), the W chromosome was used as a probe to detect the similarities among W chromosomes in the family. Zoo-FISH with W chromosome probe of *E. kuehniella* revealed a partial homology between W chromosomes for other three species. The data are consistent with phylogenetic relationships among species of Pyralidae family, and it was concluded that they have an ancient sex chromosome system with a higher molecular differentiation, which was revealed by low similarities among W chromosomes (Vítková et al. 2007).

The role of sex chromosomes in a sex phenotype development (for example, the presence of a sex-determining gene, such as in therians or medaka fish) is not broadly known in animals. Although great advances have been obtained through cytogenetics concerning sex chromosomes, several questions still remain: Why there are independent origins (multiple origins) of sex chromosomes into one group whereas other groups have monophyletic origin?; How the sex chromosomes works to define maleness or femaleness in animals? Molecular cytogenetic approaches integrated to the vast genomic data available recently may elucidate these questions and bring new insights on the biological role of sex chromosomes.

8.2. B Chromosomes

It has been reported for several genomes the presence of extra chromosomes that posses traits and different properties from most of chromosomes of the complement. These chromosomes are named B chromosomes or supernumerary chromosomes, and they are also included in the group of parasite elements or selfish element of the genome. A good example of this selfishness can be observed in the wasp *Nasonia vitripenis* species, in which the presence of a paternal sex ratio (PSR) chromosome, a kind of B chromosome, in the male gamete conducts to the loss of the complete haploid genome after the fertilization, and just the maintenance of the PSR chromosome. Thus the diploid eggs, which would develop as female, became haploid eggs and consequently, the male development occurs, which will harbor the PSR. The PSR is the most extreme known example of selfish DNA (revised by Werren and Stouthamer 2003, revised by Beukeboom et al. 2007).

Along the years, the B chromosomes have attracted the focus of studies and there is a huge amount of knowledge about these chromosomes in a range of animals, plants and fungi. Despite their broad occurrence among eukaryotes, there are still many questions remaining to be further elucidated concerning B chromosomes. In this topic, accumulated data about these chromosomes and some questions about their origin and function into the host genome will be discussed.

After the first report of occurrence of B chromosomes (Wilson 1907a, b), a huge amount of papers describing the presence of these chromosomes in a range of plants and animals was published. For example, in animal genomes, Bs in more than 500 species, including Platyhelminthes, Molusca, Arthropoda (Arachnida, Crustacea and a range of groups belonging to Insecta), fish groups, Amphibia, Reptilia, birds and mammals were reported (reviewed by Camacho 2005). Focusing on this scenario, the first question that emerges is how these chromosomes arose and evolve in such diverse genomes? Certainly, this question has been intriguing the researchers until this time, but nowadays it is possible to see some plausible explanation to this biological phenomenon.

Regarding the B chromosome origin, the aneuploidy and occurrence of other chromosomal rearrangements are the main source of their origin. Moreover, it is important that the new fragment generated after some rearrangement posseses a centromere, thus creating a neo-B. However, interestingly, the minority of these fragments will become a B chromosome, because these neo-Bs need to have an ability in maintaining themselves into the host genome. Thus, the transmission advantage (meiotic drive) or some benefits to the host genome are the most powerful mechanism to perform the maintenance. Finally, the neo-B needs to have a quick differentiation of their DNA content in an attempt to become a non-recombinant chromosome with any other A chromosomes of the host genome. Consequently, the accumulation of repetitive elements occurs (such as transposable elements), which creates the gain of heterochromatin or heterochromatin spreading or heterochromatinization process of the B chromosome (for more details see Camacho 2005). These processes create a chromosome with odd characteristics and similar traits among the different genomes.

The features that became a B chromosome, or an odd chromosome, include: their relative dispensability (a B is not present in all individuals of a species, it may be present in some cells of some individuals from some populations in some species); after their origins of A chromosomes, they follow their own evolutionary rate; they possess an irregular meiotic behavior (non-Mendelian segregation); and they, in general, do not recombine with other chromosome of the genome. In addition, another interesting trait is that the number of Bs into a genome can cause effects on the phenotype (generally, a high number of Bs is harmful while a low number may be beneficial to the host genome). Moreover, the B presence or absence contributes to intraspecific variation in the genome size (Beukeboom 1994, Jones 1991, 1995, Camacho 2005, Jones et al. 2008).

It is broadly related in the literature that the Bs have a great variability of size and morphologies, even within the same genome (for example see the review Camacho 2005). They may be the smallest chromosome of the genome, such as reported in the fly *Megaselia scalaris* (Wolf et al. 1991) and in the mouse *Reithrodontomys megalotis* (Peppers et al. 1997) or microchromosomes as noticed in the fish *Prochilotus scrofa* (Pauls and Bertollo 1983). On the other hand, the Bs may be very large, as noticed in rodent *Uromys caudimaculatus* (Baverstock et al. 1982), in the fish *Astyanax scabripinis* (Mestriner et al. 2000), *Alburnus alburnus* (the largest B of vertebrates) (Ziegler et al. 2003, Schmid et al. 2006) and *Haplochromis obliquidens* (Poletto et al. 2010b).

Despite reports on the presence of single copy genes in Bs, as noticed in several Canidae species (Graphodatsky et al. 2005; Yudkin et al. 2007), the Bs length variation are related to the DNA content of heterochromatin, which is rich in repetitive sequences such as rDNA, satellite DNA and transposable elements. Several techniques have been applied to elucidate the content of Bs, including the cytogenetic mapping of DNA sequences (single copy genes, repetitive sequences, whole chromosome painting) (see Chapter IV). In some cases, these results are useful to predict the origin of a B chromosome, which may be intraspecific (when it arises from an A chromosome of its own genome) or interspecific (when it arises from an A chromosome of another species introduced by interspecies mating). However, the precise origin of a B chromosome in a range of animals is very hard to elucidate because generally, this chromosome already suffered many rearrangements and mutations along its existence.

The first sequence used as a probe in the cytogenetic mapping in an animal, the B chromosome was a satellite DNA named 1682 of the test fly *Glossina austeni*. The results showed that the B is enriched with this sequences that is not exclusive of this chromosome.

Thus, this result helped to create the hypothesis of intraspecific origin from the B chromosome in this species (Amos and Dover 1981). After this time, several other studies using satellite DNAs as probes for *in situ* hybridization approaches were published and the intraspecific origin was also concluded, as reported in some population of *Eyprepocnemis plorans* species (Cabrero et al. 2003a), in *Drosophila subsilvestris* (originated from dot chromosomes by amplification of one satellite DNA family) (Gutknecht et al. 1995) and in the fish *Prochilodus lineatus* (using SATH1) (Jesus et al. 2003), for example. Interestingly, the results of FISH using a satellite DNA named As51 in *Astyanax scabripinnis* lead to the conclusion that the B chromosome in this species was originated through the formation of an isochromosome from a specific chromosome of the A genome (Mestriner et al. 2000).

In some cases, the intraspecific origin of B chromosomes was hypothesized based on the cytogenetic mapping of rDNA, such as observed in grasshoppers *E. plorans* (for example López-Léon et al. 1994, Cabrero et al. 2003a), *Dichroplus pratensis* (Bidau et al. 2004), *Rhammatocerus brasiliensis, Xyleus discoideus angulatus* (Loreto et al. 2008) and *Locusta migratoria* (Teruel et al. 2010), and in the fish *H. obliquidens* (Poletto et al. 2010b) and in the mammal *Rattus rattus* (Stitou et al. 2000). Other examples about the rDNA presence/absence in the Bs may be retrieved from the literature.

In a recent article (Cabral-de-Melo et al. 2010a), C_0t-1 DNA was used as a probe in FISH assays in an attempt to study B chromosomes in the beetle *Dichotomius geminatus*. The results showed that the B chromosome is enriched with repetitive elements shared with the A genome, being able to hypothesize an intraspecific origin of this chromosome and its possible process of homogeneization with A complement.

Besides the use of satDNA and multi-gene families the chromosome painting, using probes obtained from B chromosomes have lighted the origin and DNA content of this element in some species. The first B chromosome studied under painting technology in animals was the B chromosome of the marsupial *Petauroides volans*, and the results lead to propose an intraspecific origin for this chromosome (McQuade et al. 1994). This approach was also applied in the raccoon dog *Nyctereutes procyonoides* (Trifonov et al. 2002) and in the wood mouse *Apodemus peninsulae* (Trifonov et al. 2002, Karamysheva et al. 2002).

Moreover, Bs were also used as a source of DNA in cloning, sequencing and dot/southern blot analysis. This technology in B chromosome researches was applied in frog *Leiopelma hochstetteri*, showing an enrichment of repetitive elements in this chromosome (Sharbel et al. 1998), and in the salamander *Dicamptodon tenebrosus*, which reveled a sharing of repetitive sequences with genomes of closely related species (for more details see Brinkman et al. 2000). In both cases, it was also reported the putative intraspecific origins to these Bs (Sharbel et al. 1998, Brinkman et al. 2000).

In spite of many articles available in the literature, well-studied B chromosomes can be observed in Ortopthera, such as in the grasshoppers *Eyprepocnemis plorans* and *Locusta migratoria*, being that these Bs are classical cases of supernumeraries chromosomes in animal genomes. In an attempt to investigate the B chromosomes origin in both species, the chromosome painting assays using as probe the Bs of both species was performed (for *E. plorans* the B_{24}) and X chromosomes, besides other kinds of probes (such as rDNA, histone, genes, satDNAs). In *L. migratoria,* the B chromosome probe revealed that the B chromosome shares sequences with pericentromeric regions of X and most of A chromosomes. Moreover, it was shown that the X chromosome shares sequences with pericentromeric regions of two A chromosomes, some non-centromeric regions of most A chromosomes and on most parts of the B chromosome. These results suggested an intraspecific origin of B chromosomes in *L. migratoria* and, even sharing some sequences with X chromosome, it was not possible to determine the specific A chromosomes from which the B originated (Teruel et al. 2009a). Although the use of B chromosome paint did not reveal the origin of the B chromosome in *L. migratoria*, the precise origin of this element was revealed using histone genes hybridization. The B chromosome of this species has H3 and H4 histone genes that are shared with the autosomal pair 8, indicating the origin from this chromosome is likely to be more than 750,000 years old, based on the divergence time of these genes among B and A chromosomes (Teruel et al. 2010).

In the grasshopper *E. plorans,* previous studies on the B_{24} chromosome using rDNA and satelitte DNA as probes, suggested that the B chromosome is derived of the X chromosome, despite the probe sequences being present in most of the autosomes. Thus, chromosome painting using B_{24} and X chromosomes as a probe was not able to reveal if the B chromosome is derived from the X chromosome or from one or more autosomes, because the probes hybridized the whole B_{24} chromosome and also the proximal regions of all A chromosomes (Teruel et al. 2009b). However, these experiments corroborated the observation that the X chromosome and B_{24} chromosome share repetitive elements (López-León et al. 1994, Cabrero et al. 1999, Teruel et al. 2009b). Interestingly, a previous study also using rDNA and satelitte DNA as probes in populations of Eastern (Caucasus) and Western (Spain and Morocco) of this species, showed that the B chromosomes of both places were originated independently from the autosomes and X chromosome, respectively (Cabrero et al. 2003a).

The interespecific origin to Bs is also reported in the literature. For example, molecular evidences showed that the PSR in *Nasonia vitripennis* possesses a hybrid origin, which entered in this genome by hybridization with *N. longicornis*. Other results support the idea of the hybrid origin to this chromosome through a recent hybridization between *N. vitripennis* and a wasp of *Trichromalopsis* genus (revised in McAllister and Werren 1997). Interestingly, in other experiments it was observed that in the *Nasonia* genus, a neo-B chromosome is able to emerge in experimental interspecific crosses. In addition, it was possible to conclude that this chromosome emerged from chromosome 4 (Perfectti and Werren 2001). Additionally, it is suggested that the PSR in *Trichogramma* was also originated by interespefic hybridization (for more description of the papers about PSR see the review of Werren and Stouthamer 2003).

Other kind of probes have also been applied to understand the evolution of B chromosomes. For example, the telomeric probes were used in FISH assays in the raccoon dog *N. procyonides* (Wurster-Hill et al. 1988) and in the neotropical water rat *Nectomys squamipes* (Silva and Yonenaga-Yassuda 1998), showing interstitital sites in the Bs. The results suggested that chromosomal rearrangements, such as fusions or translocations, could be important to the evolution of a B chromosome (see Camacho 2005). Moreover, TEs are also used as probes, such as reported in the mammal *Reithrodontomys megalotis* (Peppers et al. 1997) (for more reports of TEs presence in the Bs see Camacho 2005).

The accumulation of huge amount of repetitive elements in the B chromosomes turns hard to trace its origin from a specific A chromosomes. The particular evolutionary dynamics of repeated DNAs lead to the spreading of copies between the A and B elementes, what make difficulty any comparative analysis. Thus, in attempt to build a more precise origin of a B chromosome it is necessary to have a molecular marker that possess any important trait, as an conservation of number and position of sites in a range of species, mainly in the species related to the "target species" (species that harbor the B); and a marker that is present in the B chromosome, even in few or degenerated copies.

The next step in the studies on Bs is the understanding the role of these chromosomes in the genome, although the studies concerning composition and origin of these elements are still important. About this question, it is important to investigate some relationship among Bs and phenotypic traits or to find genes in their DNA content. The scarce data about single copy genes into Bs does not mean that they do not posses genes in their genomic contents, instead it may be a consequence of the presence of an "ocean" of repetitive DNAs that makes it hard to find the single copy genes. The emergence of new generation sequencing and microarray, associated to the molecular cytogenetics, represents powerful techniques to find genes in B chromosomes and their possible role. The advanced technologies on molecular biology will open new windows in the direction of the application of functional genomics to B chromosomes, increasing the knowledge about the origin and evolution of these chromosomes.

8.3. B AND SEX CHROMOSOMES: SIMILAR OR DIFFERENT?

It is notorious that the B and sex chromosomes have different traits in relation to the other chromosomes of the genome, related to origin, composition, and evolution. These characteristics make these chromosomes "odd" elements with some similar, and at the same time, singular characteristics among them and among the rest of the genome.

The common traits between sex and B chromosomes include (i) meiotic univalency in some cases (as in the X0 system); (ii) similar condensation pattern along the meiosis (Camacho 2005) for some groups, for example grasshoppers; (iii) a distinct degree of recombination absence (noticed along the length of Bs and between parts of homologous of sex chromosome pairs); (iv) enrichment of repetitive elements in their chromatin; and (v) they may arise from any chromosome of their own genome (except in the cases of interspecific origin of Bs). Thus, the question is: in a hypothetical situation in which a "B chromosome" has some recombination area, even in a small part of this chromosome, or if a "sex chromosome" does not has a homologue (similar to X0 system) besides an absence of any identifiable factor that confer a sexual phenotype development, or another uncommon and hypothetical situation, these hypothetical chromosomes would be B chromosomes or sex chromosomes? Obviously, if other specific traits in one of these systems (Bs or sex chromosomes) are present, it became easy to classify them. It is interesting to think in the putative superposition of traits, origin, and function of Bs and sex chromosomes. In fact, along the earlier studies about odd chromosomes, McClung (1901, 1902) concluded that the additional chromosome present in males in an insect species (the author named this chromosome as an "accessory chromosome") is involved in sex determination in this species. Actually, this report (McClung 1901) is cited as the first discovery of a sex chromosome (see reviews available in the literature).

Moreover, there are several examples in the literature that causes us to think in the border between Bs and sex chromosomes, as exemplified by, (i) the PSR chromosome in *N. vitripennis* (discussed in the first paragraph of the previous topic). Other examples include (ii) the B chromosome origin from a sex chromosome discussed in the fly *Glossina*, the first article using the physical mapping approach to study sex chromosomes in animal genomes (Amos and Dover 1981); (iii) the achiasmatic Y (present in an original X0 system) of one population of Homoptera *Cacopsyllla peregrine*, evolved from a mitotically stable B chromosome that integrated into an achiasmatic segregation mechanism with X, thus fixed as a Y chromosome (Nokkala et al. 2003); (iv) similar results were obtained in the psyllid *Rhinocola aceris* (interestingly, in this species X and B fail to segregate in <2% of the meiosis) (Nokkala et al. 2000); (v) in *Drosophila*, it is predicted that the Y chromosome might be a specialized B chromosome (Hackstein et al. 1996, Carvalho 2002); (vi) the meiotic behavior of X and B chromosomes in the grasshopper *E. plorans* is conducive to the explaination that the Bs analyzed could be originated from X chromosomes or reflect a similar chromatin structure between them (Viera et al. 2004); (vii) the male meiotic behavior of univalents X, B and A chromosomes reveals that along the meiosis cycle, the B chromosome may behave similar to the X chromosome (Rebollo et al. 1998); (viii) in the frog *Leiopelma hochstetteri,* the sex chromosome originated a B chromosome, as a result of isolation from recombination between the homologous of a sex chromosome pair and posterior accumulation of divergence among the sequences in both chromosomes (for more details see Sharbel et al. 1998). For more details of the examples of this paragraph, see Camacho (2005), Carvalho (2002) and Carvalho et al. (2009) and references therein.

Despite these similarities between Bs and sex chromosomes, this may reflect analogy and not homology, in the cases that the B or sex chromosomes origins are not clearly elucidated. It may occur because these chromosomes have similar evolutionary processes (for example genetic degeneration) (see Camacho 2005). For example, in the beginning of this chapter it was reported that the lack of recombination and genetic degeneration are conceptually linked to sex chromosome evolution that it is associated with the current concept of sex chromosomes. Moreover, other chromosome systems, which have suppression of recombination, also show some similarities to sex chromosomes (Charlesworth et al. 2005), for example, autosomal heteromorphic chromosomes or Bs. Thus, the sex chromosomes are special cases of more general chromosome phenomenon (Fraser and Heitman 2004) and in this way, the same idea may be applied to B chromosomes.

Finally, the sex chromosomes and B chromosomes are the focus of most cytogenetic studies, however the knowledge about their origin, evolution and function needs to be elucidated with more details. In the new genomic era, in which new technologies are arising, new approaches are becoming more accessible to researchers. Consequently, the integration of new technologies of DNA sequencing, gene expression analysis, physical genome mapping, cytogenetic mapping and bioinformatics applied in the sex and B chromosomes studies will increase the knowledge, as well as give new insights about the structure, function and evolution of these chromosomes.

Chapter 9

FINAL REMARKS

The advent of molecular technologies in the 1980s suggested that cytogenetics would become extinct and the molecular tools would solve most biological questions. This statement was insinuated in the words of the USA president Bill Clinton on June 25, 2000 during the announcement of the completion of the first draft of the entire human genome acquired by the International Human Genome Project and Celera Genomics Corporation. Clinton said "**Genome science** will have a real impact on all our lives - and even more, on the lives of our children. It will revolutionize the diagnosis, prevention and treatment of most, if not all, human diseases. In coming years, doctors increasingly will be able to cure diseases like Alzheimer's, Parkinson's, diabetes and cancer by attacking their genetic roots. Just to offer one example, patients with some forms of leukemia and breast cancer already are being treated in clinical trials with sophisticated new drugs that precisely target the faulty genes and cancer cells, with little or no risk to healthy cells. In fact, it is now conceivable that our children's children will know the term cancer only as a constellation of stars". Ten years after **Clinton**'s speech, **cancer, diabetes, Alzheimer's and Parkinson's are still cau**sing huge damages in humans health and several biological questions related to the origin and evolution of most genome variation is still waiting for answers. It is clear that the advances in molecular biology bring new ways to prevent, diagnose, treat and cure disease in animals and humans, besides adding new information about the evolutionary process and the diversity of genomes. On the other hand, most recent studies have shown that deciphering the nucleotide sequence of a complete genome does not mean so much and most differences among organisms are related not to the physical structure of genes, but to the control of how the genes work involving other than changes in the underlying DNA sequence packed in chromosomes.

Important information about the genomes can be acquired from fundamental cytogenetic studies based on the identification of chromosome number and morphology, up to advanced molecular and bioinformatic approaches applied in cytogenetics. In this book, we reviewed the contribution of this area to the knowledge of animal genomes. In this way, the cytogenetics does not become extinct but is integrating the advances on genomics and molecular biology **moving to a "next generation cytogenetics". One of the most significant** contributions of cytogenetics to genomics can be illustrated by the pufferfish *Tetraodon nigroviridis*, one vertebrate organism model for genomic studies. This species contains the most compact genome of a vertebrate species with 340 Mb. Molecular cytogenetics was applied to anchor nucleotide sequence data to specific chromosomes of the species, allowing comparative analysis of other vertebrates and inferences on the ancestral bony vertebrate, which was composed of 12 chromosomes (Jaillon et al. 2004). Furthermore, analysis of the *Tetraodon* and human genomes shows that whole-genome duplication occurred in the teleost fish lineage, subsequent to its divergence from mammals.

In Chapter III of this book, we emphasized that cytogenetics has gained importance considering the integration of genomic data to chromosome analysis (integrative cytogenetics). The application of bioinformatics to generate information on the chromosome level represents an important step in this direction. The availability of complete nucleotide sequences of a large number of animal genomes brings the possibility of a new area for chromosome analysis, the *in silico* cytogenetics. The most recent view of the "integrative cytogenetics" is in the direction of analysis involving the "next generation" methods of genome sequencing, "microarrays" and "micro RNA molecules", which opens a new possibility of understanding the genomes under the focus of cytogenetics (see Figure 9.1). Whole genomes or chromosomes can be analyzed against microarray platforms looking for functional or structural patters of specific chromosomes or genomes. Such an approach seems to be very efficient to investigate several important questions in cytogenetics, such as sex chromosomes and B chromosomes. In the same way, the discovery of microRNAs and their involvement with several biological processes also represents a new focus of analysis for cytogenetics that can involve the *in silico* cytogenetic mapping of microRNAs against other genetic/genomic maps. The integration of cytogenetics and genomics emerges as a new area in biology, the "cytogenomics".

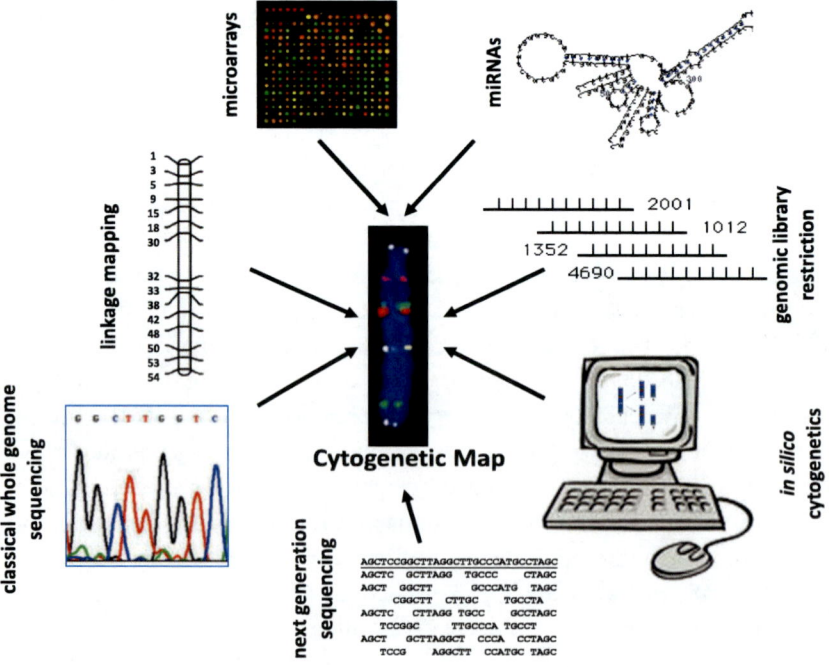

Figure 9.1. Integrative view of cytogenetics.

On the other hand, although a range of genomic technologies are available and promises fantastic contribution to the cytogenetic area, the physical analysis of the chromosomes and their visualization under a microscope constitute the most fascinating and promising way to understand genomes throughout cytogenetics.

Chapter 10

REFERENCES

Abdelaziz, M., Teruel, M., Chobanov, D., Camacho, J. P. M. and Cabrero, J. (2007). Physical mapping of rDNA and satDNA in A and B chromosomes of the grasshopper *Eyprepocnemis plorans* from a Greek population. *Cytogenet. Genome Res.,* 119, 143-146.

Abramyan, J., Ezaz, T., Graves, J. A. M. and Koopman, P. (2009). Z and W sex chromosomes in the cane toad (*Bufo marinus*). *Chromosome Res.,* 17, 1015-1024.

Abrusan, G., Krambeck, H. J., Junier, T., Giordano, J. and Warburton, P. E. (2008). Biased distributions and decay of long interspersed nuclear elements in the chicken genome. *Genetics*, 178, 573-581.

Acosta, M. J., Marchal, J. A., Mitsainas, G. P. et al. (2009). A new pericentromeric repeated DNA sequence in *Microtus thomasi. Cytogenet. Genome Res.*, 124, 27-36.

ACPAC Resource Center (BPRC) at Children's Hospital Oakland Research Institute in Oakland, California, in the United States. 2010-10-30. Available from: URL: http://bacpac.chori.org/home.htm.

Adams, K. L. and Wendel, J. F. (2005). Polyploidy and genome evolution in plants. *Curr. Opin. Plant Biol.,* 8, 135-141.

Adega, F., Chaves, R. and Guedes-Pinto, H. (2008). Suiformes orthologous satellite DNAs as a hallmark of *Pecari tajacu* and *Tayassu pecari* (Tayassuidae) evolutionary rearrangements. *Micron*, 39, 1281-1287.

Aerts, J., Crooijmans, R., Cornelissen, S. et al. (2003). Integration of chicken genomic resources to enable whole-genome sequencing. *Cytogenet. Genome Res.*, 102, 297-303.

Albig, W., Kardalinou, E., Drabent, B., Zimmer, A. and Doenecke, D. (1991). Isolation and characterization of two human H1 histone genes within clusters of core histone genes. *Genomics*, 10, 940-948.

Albig, W., Warthorst, U., Drabent, B. et al. (2003). *Mytilus edulis* core histone genes are organized in two clusters devoid of linker histone genes. *J. Mol. Evol.,* 56, 597-606.

Alkan, C., Ventura, M., Archidiacono N. et al. (2007). Organization and evolution of primate centromeric DNA from whole-genome shotgun sequence data. *PLoS Comput. Biol.*, 3, 1807-1818.

Allshire, R. C., Dempster, M. and Hastie, N. D. (1989). Human telomeres contain at least three types of G rich repeat distributed non-randomly. *Nucleic Acids Res.* 17, 4611-4627.

Almeida-Toledo, L. F. and Foresti, F. (2001). Morphologically differentiated sex chromosomes in Neotropical freshwater fish. *Genetica*, 111, 91-100.

Almeida-Toledo, L. F., Foresti, F. and Toledo-Filho, S. A. (2000). Karyotypic evolution in Neotropical freshwater fish. *Chromosomes Today*, 13, 169-182.

Almeida-Toledo, L. F., Viegas-Pequignot, E., Foresti, F., Toledo-Filho, S. A. and Dutrillaux, B. (1988). BrdU replication patterns demonstrating chromosome homoeologies in two fish species, genus *Eigenmannia*. *Cytogenet. Cell Genet.*, 48, 117-120.

Amarger, V., Gauguier, D., Yerle, M., et al. (1998). Analysis of the human, pig, and rat genomes supports a universal telomeric origin of minisatellite sequences. *Genomics*, 52, 62-71.

Amor, N., Odierna, G., Chinali, G., Said, K. and Picariello, O. (2009). Unusual Chromosomal Distribution of a Major Satellite DNA from *Discoglossus pictus* (Amphibia, Anura). *Cytogenet. Genome Res.*, 127, 33-42.

Amores, A., Force, A., Yan, Y., et al. (1998). Zebrafish *hox* clusters and vertebrate genome evolution. *Science*, 282, 1711-1714.

Amos, A. and Dover, G. (1981). The distribution of repetitive DNAs between regular and supernumerary chromosomes in species of *Glossina* (Tsetse): a two-step process in the origin of supernumeraries. *Chromosoma*, 81, 673-690.

Annunziato, A. (2008) DNA packaging: Nucleosomes and chromatin. *Nat. Educ.*, 1,1.

Ao, L., Mao, X., Nie, W. et al. (2007). Karyotypic evolution and phylogenetic relationships in the order Chiroptera as revealed by G-banding comparison and chromosome painting. *Chromosome Res.*, 15, 257-267.

Aparício, S., Chapman, J., Stupka, E. and Putnam, N. (2002). Whole-genome shotgun assembly and analysis of the genome of *Fugu rubripes*. *Science*, 297, 1301-1310.

Arnheim, N. (1983). Concerted evolution of multigene families. In M. Nei and R. K. Kochn (eds.), Evolution of genes and proteins (pp. 38-61). Sunderland: Sinauer Associates.

Ashley, T. (1994). Mammalian meiotic recombination: a reexamination. *Hum. Genet.* 94, 587-593.

Balbiani, E. G. (1881). Sur la structure du noyau des cellules salivaires chez les larves de *Chironomus*. *Zool. Anz.*, 4, 637-641.

Balloux, F., Ecoffey, E., Fumagalli, L., et al. (1998). Microsatellite conservation, polymorphism, and GC content in shrews of the genus *Sorex* (Insectivora, Mammalia). *Mol. Biol. Evol.*, 15, 473-475.

Barbazuk, W.B., Korf, I., Kadavi, C., et al. (2000). The syntenic relationship of the zebrafish and humans genomes. *Genome Res.*, 10, 1351-1358.

Baroin, A., Prat, A. and Caron, F. (1987). Telomeric site position in macronuclear DNA of *Paramecium primaurelia*. *Nucleic Acids Res.* 15, 1717-1728.

Bartolomé, C., Maside, X. and Charlesworth, B. (2002). On the abundance and distribution of transposable elements in the genome of *Drosophila melanogaster*. *Mol. Biol. Evol.*, 19, 926-937.

Barzotti, R., Pelliccia, F. and Rocchi, A. (2000). Sex chromosome differentiation revealed by genomic *in situ* hybridization. *Chromosome Res.*, 8, 459-464.

Basset, P., Yannic, G, Yang, F., et al. (2006). Chromosome localization of microsatellite markers in the shrews of the *Sorex araneus* group. *Chromosome Res.*, 14, 253-262.

Baverstock, P. R., Gelder, M. and Jahnke, A. (1982). Cytogenetic studies of the Australian rodent, *Uromys caudimaculatus*, and a species showing extensive heterochromatin variation. *Chromosoma*, 84, 517-533.

Beçak, M. L., Denaro, L. and Beçak, W. (1970). Polyploidy and mechanisms of karyotypic diversification in Amphibia. *Cytogenetics*, 9, 225-238.

Beçak, W. and Beçak, M. L. (1969). Cytotaxonomy and chromosomal evolution in Serpentes. *Cytogenetics*, 8, 247-262.

Beçak, W., Beçak, M. L., Nazareth, H. R. S. and Ohno, S. (1964). Close karyological kinship between the reptilian suborder Serpentes and the class Aves. *Chromosoma*, 15, 606-617.

Beck, T. W., Menninger, J., Voigt, G., et al. (2001). Comparative feline genomics: a BAC/PAC contig map of the major histocompatibility complex class II region. *Genomics*, 71, 282-295.

Beckmann, J. S. and Weber, J. L. (1992). Survey of human and rat microsatellites. *Genomics*, 12, 627-631.

Begun, D. J., Lindfors, H. A., Kern, A. D. and Jones, C. D. (2007). Evidence for *de novo* evolution of testis-expressed genes in the *Drosophila yakuba/Drosophila erecta* clade. *Genetics*, 176, 1131-1137.

Belancio, V. P., Hedges, D. J. and Deininger, P. (2008). Mammalian non-LTR retrotransposons: For better or worse, in sickness and in health. *Genome Res.* 18, 343-358.

Bell, G. I., De Gennaro, L. J., Gelfand, D. H., et al. (1977). Ribosomal RNA genes of *Saccharomyces cerevisiae*. I. Physical map of the repeating unit and location of the regions coding for 5S, 5.8S, 18S, and 25S ribosomal RNAs. *J. Biol. Chem.*, 252, 8118-8125.

Bernardi, G. and Bernardi, G. (1990). Compositional patterns in the nuclear genomes of cold-blooded vertebrates. *J. Mol. Evol.*, 31, 265-281.

Bertollo, L. A. C., Born, G. G., Dergam, J. A., Fenocchio, A. S. and Moreira-Filho, O. (2000). A biodiversity approach in the neotropical fish *Hoplias malabaricus*. Karyotypic survey, geographic distribution of cytotypes and cytotaxonomic considerations. *Chromosome Res.*, 8, 603-613.

Betrán, E. and Long, M. (2002). Expansion of genome coding regions by acquisition of new genes. *Genetica*, 115, 65-80.

Beukeboom, L. W. (1994). Bewildering Bs: an impression of the 1st B-Chromosome Conference. *Heredity*, 73, 328-336.

Beukeboom, L. W., Kamping, A. and van de Zande, L. (2007). Sex determination in the haplodiploid wasp *Nasonia vitripennis* (Hymenoptera: Chalcidoidea): A critical consideration of models and evidence. *Semin. Cell Dev. Biol.*, 18, 371-378.

Bi, K., Bogart, J. P. and Fu, J. (2009). An examination of intergenomic exchanges in *A. laterale*-dependent unisexual *salamanders* in the genus *Ambystoma*. *Cytogenet. Genome Res.*, 124, 44-50.

Bickham, J. W. (1984). Patterns and modes of chromosomal evolution in reptiles. In A. K. Sharma and A. Sharma (eds.), Chromosomes in evolution of eukaryotic groups (vol 2, pp. 13-40). Boca Raton: CRC Press.

Bickham, J. W., Tucker, P. K. and Legler, J. M. (1985). Diploid-triploid mosaicism: An unusual phenomenon in side-necked turtles (*Platemys platycephala*). *Science*, 227, 1591-1593.

Bidau, C. J., Rosato, M. and Martí, D. A. (2004). FISH detection of ribosomal cistrons and assortment-distortion for X and B chromosomes in *Dichroplus pratensis* (Acrididae). *Cytogenet. Genome Res.*, 106, 295-301.

Biémont, C. and Vieira, C. (2006). Genetics: Junk DNA as an evolutionary force. *Nature*, 443, 521-524.

Bigot, Y., Hamelin, M. and Periquet, G. (1990). Heterochromatin condensation and evolution of unique satellite-DNA families in two parasitic wasp species: *Diadromus pulchellus* and *Eupelmus vuilleti* (Hymenoptera). *Mol. Biol. Evol.*, 7, 351-364.

Birkham, J. W. and Hanks, B. G. (2009) Diploid-triploid mosaicism and tissue ploidy diversity within *Platemys platycephala* from Suriname. *Cytogenet. Genome Res.*, 127, 280-286.

Birstein, V. J. and Mazin, A. L. (1982). Chromosomal polymorphism of *Bufo bufo*: Karyotype and C-banding pattern of *B. b. verrucosissima*. *Genetica*, 59, 93-98.

Biscotti, M. A., Canapa, A., Olmo, E., et al. (2007). Repetitive DNA, molecular cytogenetics and genome organization in the King scallop (*Pecten maximus*). *Gene*, 406, 91-98.

Bizzaro, D., Mandrioli, M., Zanotti, M., Giusti, M. and Manicardi, G. C. (2000). Chromosome analysis and molecular characterization of highly repeated DNAs in the aphid *Acyrthosiphon pisum* (Aphididae, Hemiptera). *Genetica*, 108, 197-202.

Blackburn, E. H. (2005). Telomeres and telomerase: their mechanisms of action and the effects of altering their functions. *FEBS Lett.*, 579, 859-862.

Blackburn, E. H. and Challoner, P. B. (1984). Identification of a telomeric DNA sequences in *Tnypanosoma bnicei*. *Cell*, 36, 447-457.

Boan, F., Rodriguez, J. M., and Gomez-Marquez, J. (1998). A non-hypervariable human minisatellite strongly stimulates in vitro intramolecular homologous recombination. *J. Mol. Biol.*, 278, 499-505.

Bogart, J. P. (1973). Evolution of anuran karyotypes. In J. L. Vial (ed.), Evolutionary biology of the anurans. Contemporary research on major problems (pp. 329-349). Columbia University: Missouri Press.

Bogart, J. P. (1991). The influence of life history on karyotypic evolution in frogs. In D. M. Green and S. K. Sessions (eds.), Amphibian cytogenetics and evolution (pp. 233-255). San Diego: Academic Press.

Bogenberger, J. M., Neitzel, H. and Fittler, F. (1987). A highly repetitive DNA component common to all cervodae: its organization and chromosomal distribution during evolution. *Chromosoma*, 95, 154-161.

Böhne, A., Brunet, F., Galiana-Arnoux, D., Schultheis, C., Volff, J. N. (2008). Transposable elements as drivers of genomic and biological diversity in vertebrates. *Chromosome Res.*, 16, 203-215.

Boulesteix, M., Weiss, M. and Biémont, C. (2006). Differences in Genome Size Between Closely Related Species: The *Drosophila melanogaster* Species Subgroup. *Mol. Biol. Evol.* 23, 162-167.

Boveri, Th. (1888) Zellen Studien. Jena, Z Naturw 22:687-882.

Boveri, Th. (1909) Die Blastomerenkerne yon Ascaris megalocephala und die Theorie der Chromosomenindividualit~it. Arch Zellforsch 3:181-268.

Boveri, T. (1887). Zellen-studien. 1. Tail: Die bildung der richtungskorper bei *Ascaris megalocephala* und *Ascaris lumbricoides*. *Jenaische Zeitschrift Naturwissenschaft*, 21, 423-515.

Bowcock, A. M., Ruiz-Linares, A., Tomfohrde, J. et al. (1994). High resolution of human evolutionary trees with polimorphic microsatellites. *Nature*, 368, 455-457.

Boyle, A. L., Ballard, S. G. and Ward, D. C. (1990). Differential distribution of long and short interspersed element sequences in the mouse genome: chromosome karyotyping by fluorescence in situ hybridization. *Proc. Natl. Acad. Sci. USA*, 87, 7757-7761.

Bressa, M. J., Papeschi, A. G., Mola, L. M. and Larramendy, M. L. (1999). Meiotic studies in *Dysdercus* Guérin Méneville 1831 (Heteroptera: Pyrrhocoridae). I. Neo-XY in *Dysdercus albofasciatus* Berg 1878, a new sex chromosome determining system in Heteroptera. *Chromosome Res.*, 7, 503-508.

Bressa, M. J., Papeschi, A. G., Vítková, M. et al. (2009). Sex chromosome evolution in cotton stainers of the genus *Dysdercus* (Heteroptera: Pyrrhocoridae). *Cytogenet. Genome Res.*, 125, 292-305.

Brinkman, J. N., Sessions, S. K., Houben, A. and Green, D. M. (2000). Structure and evolution of supernumerary chromosomes in the Pacific giant salamander, *Dicamptodon tenebrosus*. *Chromosome Res.*, 8, 477-485.

Britten, R. J. and Kohne, D. E. (1968). Repeated sequences in DNA. *Science*, 161, 529-540.

Brown, S. (2008). Top billing for platypus at end of evolution tree. *Nature*, 453, 138-139.

Brown, T. A. (2002). Genomes (2nd ed, 608 p.). Oxford: Bios Scientific Publishers.

Burke, W. D., Eickbush, D. G., Xiong, Y., Jakubczak, J. L. and Eickbush, T. H. (1993). Sequence relationship of retrotransposable elements R1 and R2 within and between divergent insect species. *Mol. Biol. Evol.*, 10, 163-185.

Burke, W. D., Malik, H. S., Lathe III, W. C. and Eickbush T. H. (1998). Are retrotransposons long-term hitchhikers? *Nature*, 392, 141-142.

Burke, W. D., Singh, D. and Eickbush T. H. (2003). R5 Retrotransposons Insert into a Family of Infrequently Transcribed 28S rRNA Genes of Planaria. *Mol. Biol. Evol.*, 20, 1260-1270.

Burke, W. D., Müller, F. and Eickbush, T. H. (1995). R4, a non-LTR retrotransposon specific to the large subunit Rrna gene of nematodes. *Nucleic Acids Res.*, 23, 4628-4634.

Burt, D. W., Bruley, C., Dunn, I. C., et al. (1999). The dynamics of chromosome evolution in birds and mammals. *Nature*, 402, 411-413.

Cabral-de-Mello, D. C., Martins, C., Souza, M. J. and Moura, R. C. (2010c). Cytogenetic mapping of 5S and 18S rRNAs and H3 histone genes in four ancient Proscopiidae grasshopper species: contribution to understanding the evolutionary dynamics of multigene families. *Cytogenet. Genome Res.*, in press.

Cabral-de-Mello, D. C., Moura, R. C. and Martins, C. (2010a). Chromosomal mapping of repetitive DNAs in the beetle *Dichotomius geminatus* provides the first evidence for an association of 5S rRNA and histone H3 genes in insects, and repetitive DNA similarity between the B chromosome and A complement. *Heredity*, 104, 393-400.

Cabral-de-Mello, D. C., Moura, R. C., Carvalho, R. and Souza, M. J. (2010b). Cytogenetic analysis of two related *Deltochilum* (Coleoptera, Scarabaeidae) species: Diploid number reduction, extensive heterochromatin addition and differentiation. *Micron*, 41, 112-117.

Cabrero, J. and Camacho, J. P. M. (2008). Location and expression of ribosomal RNA genes in grasshoppers: Abundance of silent and cryptic loci. *Chromosome Res.*, 16, 595-607.

Cabrero, J., Bakkali, M., Bugrov, A., et al. (2003a). Multiregional origin of B chromosomes in the grasshopper *Eyprepocnemis plorans*. *Chromosoma*, 112, 207-211.

Cabrero, J., Bugrov, A., Warchalowska-Sliwa, E., et al. (2003b). Comparative FISH analysis in five species of *Eyprepocnemidine* grasshoppers. *Heredity*, 90, 377-381.

Cabrero, J., López-León, M. D., Bakkali, M. and Camacho, J. P. M. (1999). Common origin of B chromosome variants in the grasshopper *Eyprepocnemis plorans*. *Heredity,* 83, 435-439.

Cabrero, J., López-León, M. D., Teruel, M. And Camacho, J. P. (2009). Chromossome mapping of H3 and H4 histone gene clusters in 35 species of acridid grasshoppers. *Chromosome Res.*, 17, 397-404.

Camacho, J. P. (2005). B chromosomes. In T. R. Gregory (ed.), The Evolution of the Genome (pp. 223– 286). Sand Diego: Elsevier.

Cambridge Resource Center for Comparative Genomics and Sanger Center . CHROMHOME: CHROMosome HOmology Mapping and E-painting. 2010-10-29. Available from: URL: http://www.chromhome.org/.

Canapa, A., Cerioni, P. N., Barucca, M., Olmo, E. and Caputo, V. (2002). A centromeric satellite DNA may be involved in heterochromatin compactness in gobiid fishes. *Chromosome Res.*, 10, 297-304.

Capanna, C. (2000). Chromosome yesterday: a century of chromosome studies. *Chromosomes Today*, 13, 3-22.

Capriglione, T., Morescalchi, A., Olmo, E., et al. (1994a). Satellite DNAs heterochromatin and sex chromosomes in *Chionodraco hamatus* (Channichthyidae, Perciformes). *Polar Biol.*, 14, 285-290.

Capriglione, T., Cardone, A., Odierna, G. and Olmo, E. (1994b). Further data on the occurrence and evolution of satellite DNA families in the lacertid genome. *Chromosome Res.*, 2, 327-330.

Capriglione, T., de Santo, M. G., Odierna, G., Olmo, E. (1998). An alphoid-like satellite DNA sequence is present in the genome of a lacertid lizard. *J. Mol. Evol.*, 46, 240-244.

Caputo, V., Giovannotti, M., Cerioni, P. N., Splendiani, A. and Olmo, E. (2009). Chromosomal study of native and hatchery trouts from Italy (*Salmo trutta* complex, Salmonidae): conventional and FISH analysis. *Cytogenet. Genome Res.*, 124, 51-62.

Caputo, V., Odierna, G. and Aprea, G. (1994). A chromosomal study of *Eumeces* and *Scincus,* primitive members of the Scincidae (Reptilia, Squamata). *Boll Zool.*, 61, 155-162.

Carrano, A. V., Gray, J. W., Langlois, R. G., and Yu, L. C. (1983). Flow cytogenetics: Methodology and applications. In J. D. Rowley, and J. E. Ultmann (eds.), Chromosomes and Cancer (pp. 195-209). New York: Academic Press.

Carranza, S., Arnold, E. N., Geniez, P., Roca, J. L. and Mateo, J. A. (2008). Radiation, multiple dispersal and parallelism in Moroccan skinks, *Chalcides* and *Sphenops* (Squamata: Scincidae), with comments on *Scincus* and *Scincopus* and the age of the Sahara Desert. *Mol. Phylogenet. Evol.*, 46, 1071-1094.

Carrasco, L. A. P., Penman, D. J. and Bromage, N. (1999). Evidence for the presence of sex chromosomes in the Nile tilapia (*Oreochromis niloticus*) from synaptonemal complex analysis of XX, XY and YY genotypes. *Aquaculture*, 173, 207-218.

Carvalho, A. B. (2002). Origin and evolution of the *Drosophila* Y chromosome. *Curr. Opin. Genet. Dev.*, 12, 664-668.

Carvalho, A. B., Koerich, L. B. and Clark, A. G. (2009). Origin and evolution of Y chromosomes: *Drosophila* tales. *Trends Genet.*, 25, 270-277.

Casola, C., Lawing, A. M., Betrán, E., and Feschotte, C. (2007). PIF-like transposons are common in *Drosophila* and have been repeatedly domesticated to generate new host genes. *Mol. Biol. Evol.*, 24, 1872-1888.

Chandrasekaran, C. and Betrán, E. (2008). Origins of new genes and pseudogenes. *Nat. Educ.*, 1, 1.

Charlesworth, B., Snlegowskl, P. and Stephan, W. (1994). The evolution dynamics of repetitive DNA in eukaryotes. *Nature,* 371, 215-220.

Charlesworth, D., Charlesworth, B. and Marais, G. (2005). Steps in the evolution of heteromorphic sex chromosomes. *Heredity,* 95, 118-128.

Chaves, R., Santos, S. and Guedes-Pinto, H. (2004). Comparative analysis (Hippotragini versus caprini, Bovidae) of X-chromosome's constitutive heterochromatin by in situ restriction endonuclease digestion: X-chromosome constitutive heterochromatin evolution. *Genetica*, 121, 315-325.

Chen, R., Sodergren, E., Weinstock, G. M. and Gibbs, R. A. (2004). Dynamic building of a BAC clone tiling path for the rat genome sequencing project. *Genome Res.*, 14, 679-684.

Cheng, Y-M., Li, T-S., Hsieh, L-J. et al. (2009). Complex genomic organization of Indian muntjac centromeric DNA. *Chromosome Res.,* 17, 1051-1062.

Cheng, Z., Buell, C. R., Wing, R. A. and Jiang, J. (2002). Resolution of fluorescence *in-situ* hybridization mapping on rice mitotic prometaphase chromosomes, meiotic pachytene chromosomes and extended DNA fibers. *Chromosome Res.*, 10, 379-387.

Cheng, Z., Presting, G. C., Buell, C. R., Wing, R. A. and Jiang, J. (2001). High-Resolution pachytene chromosome mapping of bacterial artificial chromosomes anchored by genetic markers reveals the centromere location and the distribution of genetic recombination along chromosome 10 of rice. *Genetics*, 157, 1749-1757.

Chi, J. X., Huang, L., Nie, W. et al. (2005). Defining the orientation of the tandem fusions that occurred during the evolution of Indian muntjac chromosomes by BAC mapping. *Chromosoma,* 114, 167-172.

Chiang, E. F., Yan, Y. L., Tong, S. K., et al. (2001). Characterization of duplicated zebrafish cyp19 genes. *J. Exp. Zool.*, 290, 709–14.

Choi, I. G. and Kim, S. H. (2007). Global extent of horizontal gene transfer. *Proc. Natl. Acad. Sci. USA*, 104, 4489-4494.

Choo, K. H., Vissel, B., Nagy, A., Earle, E. and Kalitsis, P. (1991). A survey of the genomic distribution of alpha satellite DNA on all human chromosomes, and derivation of a new consensus sequence. *Nucleic Acids Res.*, 19, 1179-1182.

Chowdhary, B. P., Raudsepp, T., Fronicke, L. and Scherthan, H. (1998). Emerging patterns of comparative genome organization in some mammalian species as revealed by Zoo-FISH. *Genome Res.*, 8, 577-589.

Christidis, L. (1990). Chordata 3. B. Aves. Animal Cytogenetics, vol. 4 (116 p.). Berlin: Gebrüder Borntraeger.

Cinti, C., Santi, S. and Maraldi, N. M. (1993). Localization of single copy gene by PRINS technique. *Nucleic Acids Res.*, 21, 5799-5800.

Cinti, C., Stuppia, L. and Maraldi, N. M. (2002). Combined Use of PRINS and FISH in the study of the dystrophin gene. *Am. J. Med. Genet.*, 107, 115-118.

Ciobanu, D. G., Grechko, V. V. and Darevsky, I. S. (2003). Molecular evolution of satellite DNA CLsat in lizards from the genus *Darevskia* (Sauria: Lacertidae): correlation with species diversity. *Russ. J. Genet.*, 39, 1527-1541.

Ciobanu, D. G., Grechko, V. V., Darevsky, I. S. and Kramerov, D. A. (2004). New satellite DNA in *Lacerta* s. str. lizards: evolutionary pathways and phylogenetic impact. *J. Exp. Zoolog. B Mol. Dev. Evol.,* 302, 505-516.

Cioffi, M.B., Martins, C., Centofante, L., Jacobina, U., Bertollo, L.A.C. (2009). Chromosomal variability among allopatric populations of Erythrinidae fish *Hoplias malabaricus*: mapping of three classes of repetitive DNAs. *Cytogenet. Genome Res.*, 125, 132-141.

Clabby, C., Goswami, U., Flavin, F., et al. (1996). Cloning, characterization and chromosomal location of a satellite DNA from the Pacific oyster, *Crassostrea, gigas. Gene*, 168, 205-209.

Clark, A. G., Eisen, M.B., Smith, D.R., et al. (2007). Evolution of genes and genomes on the *Drosophila* phylogeny. *Nature*, 450:203-218.

Cnaani, A., Lee, B. -Y., Ozouf-Costaz, C., et al. (2007). Mapping of sox2 and sox14 in tilapia (*Oreochromis* spp.). *Sex. Dev.*, 1, 207-210.

Cnaani, A., Lee, B. -Y., Zilberman, N., et al. (2008). Genetics of sex determination in tilapiine species. *Sex. Dev.*, 2, 43-54.

Cohen, M. M. and Gans, C. (1970). The chromosomes of the order Crocodilia. *Cytogenetics*, 9, 81-105.

Colomba, M. S., Vitturi, R., Castriota, L., Bertoni, R. and Libertini, A. (2002). FISH mapping of 18S-28S and 5S ribosomal DNA, (GATA)n and (TTAGGG)n telomeric repeats in the periwinkle *Melarhaphe neritoides* (Prosobranchia, Gastropoda, Caenogastropoda). *Heredity*, 88, 381-384.

Contreras, L. C., Torres-Mura, J. C. and Spotorno, A. E. (1990). The largest known chromosome number for a mammal, in a South American desert rodent. *Experientia*, 15, 506-508.

Cook, J. M. (1993). Sex determination in the Hymenoptera – a review of models and evidence. *Heredity*, 71, 421-435.

Cordaux, R. (2008). The human genome in the LINE of fire. *Proc. Natl. Acad. Sci. USA,* 105, 19033-19034.

Cordaux, R., Udit, S., Batzer M. A. and Feschotte, C. (2006). Birth of a chimeric primate gene by capture of the transposase gene from a mobile element. *Proc. Natl. Acad. Sci. USA*, 103, 8101-8106.

Corradini, N., Rossi, F., Vern, F. and Dimitri, P. (2003). FISH analysis of *Drosophila melanogaster* heterochromatin using BACs and P elements. *Chromosoma*, 112, 26-37.

Coullin, P., Bed'Hom, B., Candelier, J. J., et al. (2005). Cytogenetic repartition of chicken CR1 sequences evidenced by PRINS in Galliformes and some other birds. *Chromosome Res.,* 13, 665-673.

Coullin, P., Roy, L., Pellestor, F., et al. (2002). PRINS, the Other *in situ* DNA Labeling Method Useful in Cellular Biology. *Amer. J. Med. Genet.*, 107, 127-135.

Crollius, H. R., Jaillon, O., Dasilva, C., et al. (2000). Characterization and repeat analysis of the compact genome of the freshwater pufferfish *Tetraodon nigroviridis. Genome Res.* 10, 939-949.

Dasilva, C., Hadji, H., Ozouf-Costaz, C., et al. (2002). Remarkable compartmentalization of transposable elements and pseudogenes in the heterochromatin of the *Tetraodon nigroviridis* genome. *Proc. Natl. Acad. Sci. USA*, 99, 1636-1641.

Dávila-Rodríguez, M. I., Cortés-Gutiérrez, E. I., López-Fernández, C., Mezzanotte, R. and Gosálvez, J. (2009). Whole-comparative genomic hybridization in domestic sheep *(Ovis aries)* breeds. *Cytogenet. Genome Res.*, 124, 19-26.

Dawid, I. B. and Rebbert, M. L. (1981). Nucleotide sequence at the boundaries between gene and insertion regions in the rDNA of *D. melanogaster*. *Nucleic Acids Res.*, 9, 5011-5020.

Dehal, P. and Boore, J. L. (2005). Two rounds of whole genome duplication in the ancestral vertebrate. *PLoS Biol.*, 3, e314.

del Gaudio, R., Potenza, N., Stefanoni, P., Chiusana, M. L. and Geraci, G. (1998). Organization and nucleotide sequence of the cluster of five histone genes in the Polichaete worm *Chaetopterus variopedatus*: first record of a H1 histone gene in the phylum Annelida. *J. Mol. Evol.*, 46, 64-73.

Demuth, J. P., De Bie, T., Stajich, J. E., Cristianini, N. and Hahn, M. W. (2006). The evolution of mammalian gene families. *PLoS One*, 1, e85.

Derjusheva, S., Kurganova, A., Krasikova, A., et al. (2003). Precise identification of chicken chromosomes in the lampbrush form using chromosome painting probes. *Chromosome Res.*, 11, 749-57.

Devlin, R. H. and Nagahama, Y. (2002). Sex determination and sex differentiation in fish: an overview of genetic, physiological, and environmental influences. *Aquaculture,* 208, 191-364.

Devlin, R. H., McNeil, B. K. and Donaldson, E. M. (1991). Isolation of a Y-chromosomal DNA probe capable of determining sex in chinook salmon. *Can J. Fish Aquat Sci.*, 48, 1606-1612.

DiNapoli, L. and Capel, B. (2008). SRY and the Standoff in Sex Determination. *Mol. Endocrinol.,* 22, 1-9.

Diniz, D., Laudicina, A., Cioffi, M. B. and Bertollo, L. A. C. (2008). Microdissection and whole chromosome painting. Improving sex chromosome analysis in *Triportheus* (Teleostei, Characiformes). *Cytogenet. Genome Res.*, 122, 163-168.

Doak, T. G., Doerder, F. P., Jahn, C. L. and Herrick, G. (1994). A proposed superfamily of transposase genes: transposon-like elements in ciliated protozoa and a common `D35E' motif. *Proc. Natl. Acad. Sci. USA,* 91, 942-946.

Domingo-Roura, X., Jacobson, H. A., and Weaver, R. F. (1997). Sex Linkage of Minisatellite Bands in Bobcats *(Felis rufus). J. Hered.*, 88, 527-531.

Driever, W., Stemple, D., Schier, A. and Solnica-Krezel, L. (1994). Zebrafish: genetic tools for studying vertebrate development. *Trends Genet.*, 10, 152-159.

Drosophila 12 Genomes Consortium (2007) Evolution of genes and genomes on the *Drosophila* phylogeny. *Nature*, 450, 203-218.

Drouin, G. and Moniz de Sá, M. (1995). The concerted evolution of 5S ribosomal genes linked to the repeated units of other multigene families. *Mol. Biol. Evol.*, 12, 481-493.

Dubcovsky, J. and Dvörak, J. (1995). Ribosomal RNA multigene loci: nomads of the Triticeae genomes. *Genetics,* 140, 1367-1377.

Dunn, K. A., Thomas, R., Binns, M. M. and Breen, M. (2000). Comparative Genomic Hybridization (CGH) in Dogs – Application to the Study of a Canine Glial Tumor. *Cell Line*, 160, 77-82.

Duret, L., Marais, G. and Biémont, C. (2000). Transposons but not retrotransposons are located preferentially in regions of high recombination rate in *Caenorhabditis elegans*. *Genetics*, 156, 1661-1669.

Eickbush, T. H. and Eickbush, D. G. (2007). Finely orchestrated movements: evolution of the ribosomal RNA genes. *Genetics*, 175, 477-485.

Eirín-López, J. M., Ruiz, M. F., González-Tizón, A. M., et al. (2004). Molecular evolutionary characterization of the mussel Mytilus histone multigene family: First record of a tandemly repeated unit of five histone genes containing an H1 subtype with "orphon" features. *J. Mol. Evol.*, 58, 131-144.

Elder, F. F. B. and Hsu, T. C. (1988). Tandem fusions in the evolution of mammalian chromosomes. In A. A. Sandberg (ed.), The Cytogenetics of Mammalian Autosomal Rearrangements (pp. 481-506). New York: Alan R Liss.

Ellegren, H. (2010). Evolutionary stasis: the stable chromosomes of birds. *Trends Ecol. Evol.*, 25, 283-291.

Ellegren, H., Johansson, M., Hartman, G. and Andersson, L. (1994). DNA fingerprinting with the human 33.6 minisatellite probe identifies sex in beavers *Castor liber*. *Mol. Ecol.*, 3, 273-274.

Endoh, H., Nagahashi, S. and Okada, N. (1990). A highly repetitive and transcribable sequence in the tortoise genome is probably a retroposon. *Eur. J. Biochem.*, 189, 25-31.

Endoh, H. and Okada, N. (1986). Total DNA transcription in vitro: a procedure to detect highly repetitive and transcribable sequences with tRNA-like structures. *Proc. Natl. Acad. Sci. USA,* 83, 251-255.

Engel, J. D. and Dodgson, J. B. (1981). Histone genes are clustered but not tandemly repeated in the chicken genome. *Proc. Natl. Acad. Sci. USA*, 78, 2856-2860.

Estoup, A., Solignac, M., Harry, M. and Cornuet, J. M. (1993). Characterization of $(GT)_n$ and $(CT)_n$ microsatellites in two insect species *Apis mellifera* and *Bombus terrestris*. *Nucleic Acids Res.*, 21, 1427-1431.

Ezaz, T., Sarre, S. D., O'Meally, D., Graves, J. A. M. and Georges, A. (2009a). Sex Chromosome Evolution in Lizards: Independent Origins and Rapid Transitions. *Cytogenet. Genome Res.*, 127, 249-260.

Ezaz, T., Moritz, B., Waters, P., et al. (2009b). The ZW sex microchromosomes of an Australian dragon lizard share no homology with those of other reptiles or birds. *Chromosome Res.,* 17, 965-973.

Ezaz, T., Quinn, A. E., Miura, I. et al. (2005). The dragon lizard *Pogona vitticeps* has ZZ/ZW microsex chromosomes. *Chromosome Res.*, 13, 763-776.

Ezaz, T., Valenzuela, N., Grützner, F., et al. (2006a). An XX/XY sex microchromosome system in a freshwater turtle, *Chelodina longicollis* (Testudines: Chelidae) with genetic sex determination. *Chromosome Res.*, 14, 139-150.

Ezaz, T., Stiglec, R., Veyrunes, F. and Graves, J. A. M. (2006b). Relationships between vertebrate ZW and XY sex chromosome systems. *Curr. Biol.,* 16, 736-743.

Feldberg, E., Porto, J. I. R. and Bertollo, L. A. C. (2003). Chromosomal changes and adaptation of cichlid fishes during evolution. In A. L. Val and B. G. Kapoor (eds.), Fish adaptations (pp. 285-308). New Dehli and New York: Science Publishers.

Ferguson-Smith, M. (2007). The Evolution of Sex Chromosomes and Sex Determination in Vertebrates and the Key Role of DMRT1. *Sex. Dev.,* 1, 2-11.

Ferguson-Smith, M. A. and Trifonov, V. (2007). Mammalian karyotype evolution. *Nature,* 8, 950-962.

Ferreira, I. A. and Martins, C. (2008). Physical chromosome mapping of repetitive DNA sequences in Nile tilapia *Oreochromis niloticus*: evidences for a differential distribution of repetitive elements in the sex chromosome. *Micron*, 39, 411-418.

Ferreira, I. A., Bertollo, L. A. C. and Martins, C. (2007). Comparative chromosome mapping of 5S rDNA and 5S *Hind*III repetitive sequences in Erythrinidae fishes (Characiformes) with emphasis on the *Hoplias malabaricus* 'species complex'. *Cytogenet. Genome Res.*, 118, 78-83.

Feschotte, C. and Pritham, E. J. (2007). DNA transposons and the evolution of eukaryotic genomes. *Ann. Rev. Genet.*, 41, 331-368.

Field, D. and Wills, C. (1998). Abundant microsatellite polymorphism in *Saccharomyces cerevisiae*, and the different distributions of microsatellites in eight prokaryotes and *S. cerevisiae*, result from strong mutation pressures and a variety of selective forces. *Proc. Nat. Acad. Sci. USA*, 1647-1652.

Fischer, C., Bouneau, L., Coutanceau, J-P., et al. (2004). Global heterochromatic colocalization of transposable elements with minisatellites in the compact genome of the pufferfish *Tetraodon nigroviridis*. *Gene*, 336, 175-183.

Fischer, C., Ozouf-Costaz, C., Roest Crollius, H., et al. (2000). Karyotype and chromosome location of characteristic tandem repeats in the pufferfish *Tetraodon nigroviridis*. *Cytogenet. Cell Genet.* 88, 50-55.

Flavel, R. B. (1986). Repetitive DNA and chromosomes. *Philos Trans. R. Soc. Lond. B Biol. Sci.*, 312, 227-242.

Flavell, A. J., Jackson, V., Iqbal, M. P., Riach, I. and Waddell, S. (1995). Ty1- *copia* group retrotransposon sequences in Amphibia and Reptilia. *Mol. Gen. Genet.*, 246, 65-71.

Flemming, W. (1882). Zellsubstanz, kern und zelltheilung. (424 p). Leipzig: F.C.W. Vogel.

Foresti, F., Oliveira, C., Galetti Jr, P.M. and Almeida-Toledo, L. F. (1993). Sinapto nemal complex analyses in spermatocytes of tilapia, *Oreochromis niloticus* (Pisces, Cichlidae). *Genome*, 36, 1124-1128.

Forney, J., Henderson, E. R. and Blackburn, E. H. (1987). Identification of the telomeric sequence of the acellular slime molds *Didvmium iris* and *Phvsamm polvcephalum*. *Nucleic Acids Res.* 15, 9143-9162.

Forozan, F., Karhu, R., Kononen, J., Kallioniemi, A. and Kallioniemi, O. P. (1997). Genome screening by comparative genomic hybridization. *Trends Genet.*, 13, 405-409.

Forster, A. C., McInnes, J. L., Skingle, D. C. and Symons, R. H. (1985). Non-radioactive hybridization probes prepared by the chemical labeling of DNA and RNA with a novel reagent, photobiotin. *Nucleic Acids Res.*, 13, 745-761.

Fouché, N., Özgur, S., Roy, D. and Griffith, J. D. (2006). Replication fork regression in repetitive DNAs. *Nucleic Acids Res.*, 34, 6239-6248.

Franco, F. F., Sene, F. M., and Manfrin, M. H. (2010). Low Satellite DNA Variability in Natural Populations of *Drosophila antonietae* Involved in Different Evolutionary Events. *J. Hered*, 101, 650-656.

Fransz, P. F., Armstrong, S., de Jong, J. H., et al. (2000). Integrated cytogenetic map of chromosome arm 4S of *A. thaliana*: structural organization of heterochromatic knob and centromere region. *Cell*, 100, 367-76.

Fransz, P., De Jong, J. H., Lysak, M., Castiglione, M. R. and Schubert, I. (2002). Interphase chromosomes in *Arabidopsis* are organized as well-defined chromocenters from which euchromatin loops emanate. *Proc. Natl. Acad. Sci. USA*, **99**, 14584-14589.

Fraser, J. A. and Heitman, J. (2004). Evolution of fungal sex chromosomes. *Mol. Microbiol.,* 51, 299-306.

Fraser, J. A. and Heitman, J. (2005). Chromosomal sex-determining regions in animals, plants and fungi. *Curr. Opin. Genet. Devel.,* 15, 645-651.

Freeman, J. L. and Rayburn, A. L. (2005). Localization of repetitive DNA sequences on in vitro *Xenopus laevis* chromosomes by primed *in situ* labeling (PRINS). *J. Hered.,* 96, 603-606.

Freeman, J. L., Adeniyi, A., Banerjee, R., et al. (2007). Definition of the zebrafish genome using flow cytometry and cytogenetic mapping. *BMC Genomics,* 8, 195.

Fridolfsson, A-K., Cheng, H., Copeland, N. G., et al. (1998) Evolution of the avian sex chromosomes from an ancestral pair of autosomes. *Proc. Natl. Acad. Sci. USA,* 95, 8147-8152.

Froenicke, L. (2005). Origins of primate chromosomes-as delineated by Zoo-FISH and alignments of human and mouse draft genome sequences. *Cytogenet. Genome Res.,* 108, 122-138.

Froenicke, L., Anderson, L. K., Wienberg, J. and Ashley, T. (2002). Male mouse recombination maps for each autosome identified by chromosome painting. *Am. J. Hum. Genet.,* 71, 1353-1368.

Froenicke, L., Caldés, M. G., Graphodatsky, A., et al. (2006). Are molecular cytogenetics and bioinformatics suggesting diverging models of ancestral mammalian genomes? *Genome Res.,* 16, 306-310.

Frönicke, L., Wienberg, J., Stone, G., Adams, L. and Stanyon, R. (2003). Towards the delineation of the ancestral eutherian genome organization: comparative genome maps of human and the African elephant (*Loxodonta africana*) generated by chromosome painting. *Proc. R. Soc. Lond. B Biol. Sci.,* 270, 1331-1340.

Fuková, I., Traut, W., Vítková, M., et al. (2007). Probing the W chromosome of the codling moth, *Cydia pomonella*, with sequences from microdissected sex chromatin. *Chromosoma,* 116, 135-145.

Galetti, P. M. Jr., Aguilar, C. T. and Molina, W. F. (2000). An overview of marine fish cytogenetics. *Hydrobiologia,* 420, 55-62.

Galetti, P. M. Jr., Molina, W. F., Affonso, P. R. and Aguilar, C. T. (2006). Assessing genetic diversity of Brazilian reef fishes by chromosomal and DNA markers. *Genetica,* 126, 161-177.

Galkina, S., Deryusheva, S., Fillon, V., et al. (2006). FISH on avian lampbrush chromosomes produces higher resolution gene mapping. *Genetica,* 128, 241-51.

Galkina, S., Lukina, N., Zakharova, K. and Rodionov, A. V. (2005). Interstitial (TTAGGG)(n) sequences are not hot spots of recombination in the chicken lampbrush macrochromosomes 1-3. *Chromosome Res.,* 13, 551-557.

Gall, J. G. and Pardue, M. L. (1969) Formation and detection of RNA-DNA hybrid molecules in cytological preparations. *Proc. Natl. Acad. Sci. USA,* 63, 378-383.

Gallardo, M. H., Bickham, J. W., Honeycutt, R. L., Ojeda, R. A. and Köhler, N. (1999). Discovery of tetraploidy in a mammal. *Nature,* 401, 341.

Gallardo, M. H., González, C. A. and Cebrián, I. (2006). Molecular cytogenetics and allotetraploidy in the red vizcacha rat, *Tympanoctomys barrerae* (Rodentia, Octodontidae). *Genomics,* 88, 214-221.

Gates, M. A., Kim, L., Egan, E. S., et al. (1999). A genetic linkage map for zebrafish: comparative analysis and localization of genes and expressed sequences. *Genome Res.*, 9, 334-347.

Gibson, G. and Muse, S. V. (2009). A Primer of Genome Science (3rd ed, 344 p.). Sunderland: Sinauer Associateds.

Gilbert, C., O'Brien, P. C., Bronner, G., et al. (2006). Chromosome painting and molecular dating indicate a low rate of chromosome evolution in golden moles (Mammalia, Chrysochloridae). *Chromosome Res.*, 14, 793-803.

Gilbert, W. (1978). Why genes in pieces? *Nature*, 271, 501.

Gillemans, N., McMorrow, T., Tewari, R., et al. (2003). Functional and comparative analysis of globin loci in pufferfish and humans. *Blood*, 101, 2842-2849.

Giovannoti, M., Cerioni, P. N., Caputo, V. and Olmo, E. (2009). Characterisation of a GC-Rich Telomeric Satellite DNA in *Eumeces schneideri* Daudin (Reptilia, Scincidae). *Cytogenet. Genome Res.*, 125, 272-278.

Givan, A. L. (2001). Principles of flow cytometry: an overview. In Z. Darzynkiewicz, H. A. Crissman, and J. P. Robinson (eds.), Flow Cytometry, Part A (3rd ed., pp. 19-50). San Diego: Academic Press.

Goday, C. and Esteban, M. R. (2001). Chromosome elimination in sciarid flies. *Bioessays*, 23, 242-250.

Gogolevsky, K. P., Vassetzky, N. S. and Kramerov, D. A. (2008). Bov-B-mobilized SINEs in vertebrate genomes. *Gene*, 407, 75-85.

Goldammer, T., Brunner, R. M., Rebl, A., et al. (2009). A high-resolution radiation hybrid map of sheep chromosome X and comparison with human and cattle. *Cytogenet. Genome Res.*, 125, 40-45.

Goll, M. G. and Bestor, T. H. (2002). Histone modification and replacement in chromatin activation. *Genes Dev.*, 16, 1739-1742.

Gondo, Y., Okada, T., Matsuyama, N., et al. (1998). Human Megasatellite DNA RS447: Copy-Number Polymorphisms and Interspecies Conservation. *Genomics*, 54, 39-49.

Gong, S., Zheng, C., Doughty, M. L., et al. (2003). A gene expression atlas of the central nervous system based on bacterial artificial chromosomes. *Nature*, 425, 917-925.

Gorman, G. C. (1973). The chromosomes of the Reptilia, a cytotaxonomic interpretation. In Chiarelli, A. B. and Cappana, E. (eds.), Cytotaxonomy and vertebrate evolution (pp. 349-424). London: Academic Press.

Gornung, E., Volleth, M., Capanna, E. and Castilia, R. (2008). Comparative cytogenetics of moles (Eulipotyphla, Talpidae): chromosomal differences in *Talpa romana* and *T. europaea*. *Cytogenet. Genome Res.*, 121, 249-254.

Gosden, J. and Lawson, D. (1994). Rapid chromosome identification by oligonucleotide-primed in situ DNA synthesis (PRINS). *Hum. Mol. Genet.*, 3, 931-936.

Gosden, J., Hanratty, D., Starling, J., et al. (1991). Oligonucleotide-primed in situ DNA synthesis (PRINS): a method for chromosome mapping, banding, and investigation of sequence organization. *Cytogenet. Cell Genet.*, 57, 100-104.

Graphodatsky, A. S. (2007). Comparative Chromosomics. *Mol. Biol.*, 41, 361-375.

Graphodatsky, A. S., Kukekova, A. V., Yudkin, D. V., et al. (2005). The proto-oncogene C-KIT maps to canid B-chromosomes. *Chromosome Res.*, 13, 113-122.

Graphodatsky, A. S., Yang, F., Perelman, P. L., et al. (2002). Comparative molecular cytogenetic studies in the order Carnivora: mapping chromosomal rearrangements onto the phylogenetic tree. *Cytogenet. Genome Res.,* 96, 137-145.

Graves, R. A., Marzluff, W. F., Giebelhaus, D. H. and Schultz, G. A. (1985). Quantitative and qualitative changes in histone gene expression during early mouse embryo development. *Proc. Natl. Acad. Sci. USA,* 82, 5685-5689.

Graves, J. A. M. (2008). Weird animal genomes and the evolution of vertebrate sex and sex chromosomes. *Annu. Rev. Genet.*, 42, 565-586.

Graves, J. A. M. and Peichel, C.L. (2010). Are homologies in vertebrate sex determination due to shared ancestry or to limited options? *Genome Biol.,* 11, 205.

Graves, J. A. M. and Shetty, S. (2000). Comparative genomics of vertebrates and the evolution of sex chromosomes. In M. S. Clark (ed.), Comparative Genomics (pp. 153-205). London: Kluwer.

Graves, J. A. M. and Shetty, S. (2001). Sex from W to Z: evolution of vertebrate sex chromosomes and sex determining genes. *J. Exp. Zool.*, 290, 449-462.

Grechko, V. V., Ciobanu, D. G., Darevsky, I. S. and Kramerov, D. A. (2005). Satellite DNA of lizards of the genus *Lacerta* s. str. (the group *L. agilis*), the family Lacertidae. *Dokl. Biochem. Biophys.*, 400, 44-47.

Green, D. M. (1988). Cytogenetics of the endemic New Zealand frog, *Leiopelma hochsteteri*: Extraordinary supernumerary chromosome variation and a unique sex-chromosome system. *Chromosoma*, 97, 55-70.

Green, D. M. and Sessions, S. K. (1991). Amphibian Cytogenetics and Evolution. San Diego: Academic Press.

Green, D. M. and Sessions, S. K. (2007). Karyology and cytogenetics. In H. Heatwole (ed.), Amphibian Biology (pp. 2757-2842). Chipping Norton Australia: Surrey Beatty and Sons.

Gregory, T. R. (2005a). Synergy between sequence and size in large-scale genomics. *Nat. Rev. Genet.*, 6, 699-708.

Gregory, T. R. (2005b). Genome size evolution in animals. In Gregory T. R. (ed.), The Evolution of the Genome (pp. 3-87). San Diego: Elsevier.

Gregory, T. R. (2010). Animal Genome Size Database. 2010-10-30. Vailable from: URL: http://www.genomesize.com.

Gregory, T. R. and Mable, B. K. (2005). Polyploidy in animals. In T. R. Gregory *(Ed.), The Evolution of the Genome* (pp. 427-517). San Diego: Elsevier.

Grewal, S. I. S. and Jia, S. (2007). Complexities of heterochromatin in fungi, ciliates, plants and mammals. *Nature Rev. Genet.,* 8, 35-46.

Griffin, D. K., Haberman, F., Masabanda, J., et al. (1999). Micro- and macrochromosome paints generated by flow cytometry and microdissection: tools for mapping the chicken genome. *Cytogenet. Cell Genet.*, 87, 278-281.

Griffin, D. K., Harvey, S. C., Campos-Ramos, R., et al. (2002). Early origins of the X and Y chromosomes: lessons from tilapia. *Cytogenet. Genome Res.*, 99, 157-163.

Griffin, D. K., Robertson, L. B. W., Tempest, H. G. and Skinner, B. M. (2007). The evolution of the avian genome as revealed by comparative molecular cytogenetics. *Cytogenet. Genome Res.*, 117, 64-77.

Griffin, D. K., Robertson, L. B., Tempest, H. G., et al. (2008). Whole genome comparative studies between chicken and turkey and their implications for avian genome evolution. *BMC Genomics*, 9, 168.

Grützner, F., Rens, W., Tsend-Ayush, E., et al. (2004). In the platypus, a meiotic chain of ten sex chromosomes shares genes with the bird Z and mammal X chromosomes. *Nature*, 432, 913-917.

Guan, X. Y., Meltzer, P. S. and Trent, J. M. (1994). Rapid generation of whole chromosome painting probes (WCPs) by chromosome microdissection. *Genomics*, 22, 107-107.

Guan, X. Y., Meltzer, P. S., Cao, J. and Trent, J. M. (1992). Rapid generation of region-specific genomic clones by chromosome microdissection: isolation of DNA from a region frequently deleted in malignant melanoma. *Genomics*, 14, 680-684.

Guo, X., DeBrosse, G. and Allen Jr., S. K. (1996). All-triploid Pacific oysters (*Crassostrea gigas* Thunberg) produced by mating tetraploids and diploids. *Aquaculture*, 142, 149-161.

Gutknecht, J., Sperlich, D. and Bachmann, L. (1995). A species specific satellite DNA family of *Drosophila subsilvestris* appearing predominantly in B chromosomes. *Chromosoma*, 103, 539-544.

Guttenbach, M., Nanda, I., Feichtinger, W., et al. (2003). Comparative chromosome painting of chicken autosomal paints 1-9 in nine different bird species. *Cytogenet. Genome Res.*, 103, 173-184.

Haaf, T. and Ward, D. C. (1994a). High-resolution ordering of YAC contigs using extended chromatin and chromosomes. *Hum. Mol. Genet.*, 3, 629-633.

Haaf, T. and Ward, D. C. (1994b). Structural analysis of m-satellite DNA and centromere proteins using extended chromatin and chromosomes. *Hum. Mol. Genet.*, 3, 697-709.

Hackstein, J. H. P., Hochstein, R., Hausteck-Jungen, E. and Beukeboom, L. W. (1996). Is the Y chromosome of *Drosophila* an evolved supernumerary chromosome? *BioEssays*, 18, 317-323.

Haeusler, R. A. and Engelke, D. R. (2006). Spatial organization of transcription by RNA polymerase III. *Nucleic Acids Res.* 34, 4826-4836.

Hahn, M., Han, M. V. and Han, S. G. (2007a). Gene family evolution across 12 *Drosophila* genomes. *PLoS Genetics*, 3, e197.

Hahn, M. W., Demuth, J. P. and Han, S. G. (2007b). Accelerated rate of gene gain and loss in primates. *Genetics*, 177, 1941-1949.

Hancock, J. M. (1999). Microsatellites and other simple sequences: genomic context and mutational mechanisms. In D. B. Goldtein and C. Schlotterer (eds), Microsatellites (pp 1-9). Oxford: Oxford University Press.

Hankeln, T., Keyl, H. G., Ross, R. and Schmidt, E. R. (1993). Evolution of histone gene loci in chironomid midges. *Genome*, 36, 852-862.

Harvey, S. C., Kwon, J. Y. and Penman, D. J. (2003a). Physical mapping of the brain and ovarian aromatase genes in the Nile Tilapia, *Oreochromis niloticus*, by fluorescence *in situ* hybridization. *Animal Genet.*, 34, 62-64.

Harvey, S. C., Boonphakdee, C., Campos-Ramos, R., et al. (2003b). Analysis of repetitive DNA sequences in the sex chromosomes of *Oreochromis niloticus*. *Cytogenet. Genome Res.*, 101, 314-319.

Harvey, S. C., Masabanda, J., Carrasco, L. A. P., et al. (2002). Molecular-cytogenetic analysis reveals sequence differences between the sex chromosomes of *Oreochromis niloticus*:

evidence for an early stage of sex chromosome differentiation. *Cytogenet. Genome Res.*, 79, 76-80.

Hasterok, R., Dulawa, J., Jenkins, G., Leggett, M. and Langdon, T. (2006). Multi-substrate chromosome preparations for high throughput comparative FISH. *BMC Biotechnol.*, 6, 20.

Heimpel, G. E. and Boer, J. G. (2008). Sex determination in the Hymenoptera. *Annu. Rev. Entomol.*, 53, 209-230.

Henikoff, S., Ahmad, K. and Malik, H. S. (2001). The centromere paradox: stable inheritance with rapidly evolving DNA. *Science*, 293, 1098-1102.

Henning, F., Trifonov, V., Ferguson-Smith, M. A. and Almeida-Toledo, L. F. (2008). Non-homologous sex chromosomes in two species of the genus *Eigenmannia* (Teleostei: Gymnotiformes). *Cytogenet. Genome Res.*, 121, 55-58.

Hernandez-Hernandez, A., Rincon-Arano, H., Recillas-Targa, F., et al. (2008). Differential distribution and association of repeat DNA sequences in the lateral element of the synaptonemal complex in rat spermatocytes. *Chromosoma*, 117, 77-87.

Hewitt, G. M. (1979). Grasshoppers and Crickets. Berlin: Gebrüder-Borntraeger.

Holland, P. W. H. (1999). The effect of gene duplication on homology. In G. R. Bock, and G. Cardew, G. (eds.), Homology (pp. 226-242). New Jersey: JohnWiley & Sons.

Hori, T., Suzuki, Y., Solovei, I., et al. (1996). Characterization of DNA sequences constituting the terminal heterochromatin of the chicken Z chromosome. *Chromosome Res.*, 4, 411-426.

Hornak, M., Hulinska, P., Musilova, P., Kubickova, S. and Rubes, J. (2009). Investigation of Chromosome Aneuploidies in Early Porcine Embryos Using Comparative Genomic Hybridization. *Cytogenet. Genome Res.*, 126, 210-216.

Hoskins, R. A., Nelson, C. R., Berman, B. P., et al. (2000). A BAC-Based Physical Map of the Major Autosomes of *Drosophila melanogaster*. *Science*, 287, 2271-2274.

Hotopp, J., Clark, M. E., Oliveira, D. C. S. G., et al. (2007). Widespread lateral gene transfer from intracellular bacteria to multicellular eukaryotes. *Science*, 317, 1753-1756.

Houseal, T. W., Cook, J. A., Modi, W. S. and Hale, D. W. (1995). Identification of highly conserved loci by genome painting. *Chromosome Res.*, 3, 175-181.

Hsu, J. and Benirschke, K. (1967). An atlas of mammalian chromosomes. New York: Springer-Verlag.

Hsu, T. C., Pathak, S. and Chen, T. R. (1975). The possibility of latent centromeres and a proposed nomenclature for total chromosome and whole arm translocations. *Cytogenet. Cell Genet.*, 15, 41-49.

Huan, P., Zhang, X. J., Zhang, X. J., et al. (2009). Chromosomal localization and development of SNP markers of a serine protease gene in Farrer's scallop (*Chlamys farreri*). *Yi Chuan*, 31, 1241-1247.

Huang, X., Hu, J., Zhang, C., et al. (2007). Cytogenetic characterization of the bay scallop, *Agropecten irradians irradians*, by multiple staining techniques and fluorescence *in situ* hybridization. *Genes Genet. Syst.*, 82, 257-263.

Hua-Van, A., Le Rouzic, A., Maisonhaute, C. and Capy, P. (2005). Abundance, distribution and dynamics of retrotransposable elements and transposons: similarities and differences. *Cytogenet. Genome Res.*, 110, 426-440.

Huber, I., Walter, L., Wimmer, R., et al. (2003). Cytogenetic mapping and orientation of the rhesus macaque MHC. *Cytogenet. Genome Res.*, 103, 144-149.

Hughes, A. L. and Piontkivska H. (2005). DNA repeat arrays in chicken and human genomes and the adaptive evolution of avian genome size. *BMC Evol. Biol.*, 5, 12.

Hutchison, N. (1987). Lampbrush chromosomes of the chicken, *Gallus domesticus*. *J. Cell Biol.*, 105, 1493-500.

Iannuzzi, L., Di Meo, G. P., Perucatti, A., et al. (2003). A Comparative FISH-mapping of the survival of motor neuron gene (SMN) in domestic bovids. *Cytogenet. Genome Res.*, 102, 39-41.

in den Bosch, H. A. J., Odierna, G., Aprea, G., et al. (2003). Karyological and genetic variation in Middle Eastern lacertid lizards, *Lacerta laevis* and *the Lacerta kulzeri* complex: a case of chromosomal allopatric speciation. *Chromosome Res.*, 11, 165-178.

Insua, A., López-Piñón, M. G., Feire, R. and Méndez, J. (2006). Karyotype and chromosomal location of 18S-28S and 5S ribosomal DNA in the Scallops *Pecten maximus* and *Mimachlamys varia* (Bivalvia: Pectnidae). *Genetica*, 126, 291-301.

International Chicken Genome Sequencing Consortium (2004). *Nature*, 432, 695.

International Human Genome Sequencing Consortium (2001). *Nature*, 409, 860.

Ivics, Z., Hackett, P. B., Plasterk, R. H. and Izsvak, Z. (1997). Molecular reconstruction of Sleeping Beauty, a Tc1-like transposon from fish, and its transposition in human cells. *Cell* 4, 501-510.

Jaillon, O., Aury, J-M., Brunet, F., et al. (2004). Genome duplication in the teleost fish *Tetraodon nigroviridis* reveals the early vertebrate proto-karyotype. *Nature*, 431, 947-957.

Jakubczak, J. L., Burke, W. D. and Eickbush, T. H. (1991). Retrotransposable elements Rl and R2 interrupt the rRNA genes of most insects. *Proc. Natl. Acad. Sci. USA*, 88, 3295-3299.

Janousek, B. and Mrackova, M. (2010). Sex chromosomes and sex determination pathway dynamics in plant and animal models. *Biol. J. Linnean Soc.*, 100, 737-752.

Jauch, A., Wienberg, J., Stanyon, R., et al. (1992). Reconstruction of genomic rearrangements in great apes and gibbons by chromosome painting. *Proc. Natl. Acad. Sci. USA*, 89, 8611-8615.

Jeffreys, A. J., Wilson, V. and Thein, S. L. (1985). Hypervariable 'minisatellite' regions in human DNA. *Nature*, 314, 67-73.

Jesus, C. M., Galetti Jr., P. M., Valentini, S. R. and Moreira-Filho, O. (2003). Molecular characterization and chromosomal location of two families of satellite DNA in *Prochilodus lineatus* (Pisces, Prochilodontidae), a species with B chromosomes. *Genetica*, 118, 25-32.

Jiang, J. and Gill, B. S. (2006). Current status and the future of fluorescence in situ hybridization (FISH) in plant genome research. *Genome*, 49, 1057-1068.

Jiang, J., Gill, B. S., Wangt, G-L., Ronald, P. C. and Wardt, D. C. (1995). Metaphase and interphase fluorescence in situ hybridization mapping of the rice genome with bacterial artificial chromosomes (physical mapping/genomic DNA clones). *Proc. Natl. Acad. Sci. USA*, 92, 4487-4491.

John, B. and Miklos, G. L. (1979). Functional aspects of satellite DNA and heterochromatin. In G. H. Bourne and J. F. Danielli (ed.), International Review of Cytology (vol 58, pp. 1-114). New York: Academic Press.

Jones, R. N. (1991). B-chromosome drive. *Am. Nat.*, 137, 430-442.

Jones, R. N. (1995). B chromosomes in plants. *New Phytol.*, 131, 411-434.

Jones, R. N., González-Sánchez, M., González-García, M., Vega, J. M. and Puertas, M. J. (2008). Chromosomes with a life of their own. *Cytogenet. Genome Res.*, 120, 265-280.

Juan, C., Pons, J. and Petitpierre, E. (1993). Localization of tandemly repeat DNA sequences in beetle chromosomes by fluorescent *in situ* hybridization. *Chromosome Res.*, 1, 167-174.

Kadandale, J. S., Wachtel, S. S., Tunca, Y., et al. (2000). Localization of SRY by Primed *in situ* Labeling in XX and XY Sex Reversal. *Am. J. Med. Genet.*, 95, 71-74.

Kallioniemi, A., Kallioniemi, O., Sudar, D., et al. (1992). Comparative genomic hybridization for molecular cytogenetic analysis of solid tumors. *Science*, 258, 818-821.

Kaminker, J. S., Bergman, C. M., Kronmiller, B., et al. (2002). The transposable elements of the *Drosophila melanogaster* euchromatin: a genomics perspective. *Genome Biol.*, 3, 84.1-84.20.

Kapitonov, V. V. and Jurka, J. (1999). Molecular paleontology of transposable elements from *Arabidopsis thaliana*. *Genetica*, 107, 27-37.

Kapitonov, V. V. and Jurka, J. (2001), Rolling-circle transposons in eukaryotes. *Proc. Natl. Acad. Sci. USA*, 98, 8714-8719.

Kapitonov, V. V. and Jurka, J. (2003). Molecular paleontology of transposable elements in the *Drosophila melanogaster* genome. *Proc. Natl. Acad. Sci. USA*, 100, 6569-6574.

Kapitonov, V. V. and Jurka, J. (2008). A universal classification of eukaryotic transposable elements implemented in Repbase. *Nat. Rev. Genet.*, 9, 411-412.

Karamysheva, T. V., Andreenkova, O. V., Bochkaerev, M. N., et al. (2002). B chromosomes of Korean field mouse *Apodemus peninsulae* (Rodentia, Murinae) analyzed by microdissection and FISH. *Cytogenet. Genome Res.*, 96, 154-160.

Kartavtseva, I. V. and Roslik, G. V. (2004). A complex B chromosome system in the Korean field mouse, *Apodemus peninsulae*. *Cytogenet. Genome Res.*, 106, 271-278.

Kasahara, M., Naruse, K., Sasaki S., et al. (2007). The medaka draft genome and insights into vertebrate genome evolution. *Nature*, 447, 7145-7149.

Kashi, Y., Iraqi, F., Tlkochlnsld, Y., et al. (1990). (TG). uncovers a sex-specific hybridization pattern in cattle. *Genomics*, 7, 31-36.

Kasowski, M., Grubert, F., Heffelfinger, C., Hariharan, M. et al. (2010). Variation in Transcription Factor Binding Among Humans. *Science*, 328, 232-235.

Katagiri, T., Kidd, C., Tomasino, E., et al. (2005). A BAC-based physical map of the Nile tilapia genome. *BMC Genomics*, 6, 89, 1-6.

Kato, A., Vega, J. M., Han, F., Lamb, J. C. and Birchler, J. A. (2005). Advances in plant chromosome identification and cytogenetic techniques. *Curr. Opinion Plant Biol.*, 8, 148-154.

Katzen, A. L., Cann, G. M. and Blackburn, E. H. (1981). Sequence specific fragmentation of macronuclear DNA in a holotrichous ciliate. *Cell* 24, 313-320.

Kawagoshi, T., Nishida, C., Ota, H. et al. (2008). Molecular structures of centromeric heterochromatin and karyotypic evolution in the Siamese crocodile (*Crocodylus siamensis*) (Crocodylidae, Crocodylia). *Chromosome Research*, 16, 1119-1132.

Kawai, A., Ishijima, J., Nishida, C. et al. (2009). The ZW sex chromosomes of *Gekko hokouensis* (Gekkonidae, Squamata) represent highly conserved homology with those of avian species. *Chromosoma*, 118, 43-51.

Kawai, A., Nishida-Umehara, C., Ishijima, J. et al. (2007). Different origins of bird and reptile sex chromosomes inferred from comparative mapping of chicken Z linked genes. *Cytogenet. Genome Res.*, 117, 92-102

Kawakami, K. (2007). *Tol2*: a versatile gene transfer vector in vertebrates. *Genome Biology*, 8, S7.

Kazazian, H. H. (2004). Mobile elements: drivers of genome evolution. *Science*, 303, 1626-1632.

Kedes, L. H. (1979). Histone genes and histone messengers. *Ann. Rev. Biochem.*, 48, 837-870.

Kedes, L. H. and Gross, P. (1969). Identification in cleaving embryos of three RNA species serving as templates for the synthesis of nuclear proteins. *Nature*, 223, 1335-1339.

Kellogg, M. E., Burkett, S., Dennis, T. R. et al. (2007). Chromosome painting in the manatee supports Afrotheria and Paenungulata. *BMC Evol. Biol.*, 7, 6.

Kerrebrock, A. W., Srivastava, R. and Gerbi, S. A. (1989). Isolation and characterization of ribosomal DNA variants from *Sciara coprophila*. *J. Mol. Biol.*, 210, 1-13.

Kidwell, M. G. (2002). Transposable elements and the evolution of genome size in eukaryotes. *Genetica*, 115, 49-63.

Kidwell, M. G. (2005). Transposable elements. In T. R. Gregory (ed.), The evolution of the genome (pp. 165-221). San Diego: Elsevier.

King, L. M. and Cummings, M. P. (1997). Satellite DNA repeat sequence variation is low in three species of burying beetles in the genus *Nicrophorus* (Coleoptera: Silphidae). *Mol. Biol. Evol.*, 14, 1088-1095.

King, M. (1990). Chordata 2. Amphibia. In: B. John and C. Gwent (eds.), Animal Cytogenetics (241 p.). Berlin: Gebrüder Borntraeger.

King, M. (1993). Species evolution, the role of chromosome change (301 p.). London: Cambridge University.

Kirby, P., Greaves, I. K., Koina, E., Waters, P. D. and Graves, J. A. M. (2007). Core-SINE blocks comprise a large fraction of monotreme genomes; implications for vertebrate chromosome evolution. *Chromosome Res.*, 15, 975-984.

Koch, J., Hindkjaer, J., Mogensen, J., Kolvraa, S. and Bolund, L. (1991). An improved method for chromosome-specific labeling of alpha satellite DNA in situ by using denatured double-stranded DNA probes as primers in a primed in situ labeling (PRINS) procedure. *Genet. Anal. Tech. Appl.*, 8, 171-178.

Kohn, M., Högel, J., Vogel, W. and et al. (2006). Reconstruction of a 450-My-old ancestral vertebrate protokaryotype. *Trends Genet.*, 22, 203-210.

Kojima, K. K. and Fujiwara, H. (2005). Long-term inheritance of the 28S rDNA-specific retrotransposon R2. *Mol. Biol. Evol.*, 22, 2157-2165.

Kojima, K. K., Kuma, K., Toh, H. and Fujiwara, H. (2006). Identification of rDNA-specific non-LTR retrotransposons in Cnidaria. *Mol. Biol. Evol.*, 23, 1984-1993.

Kojima, N. F., Kojima, K. K., Kobayakawa, S. et al. (2010). Whole chromosome elimination and chromosome terminus elimination both contribute to somatic differentiation in Taiwanese hagfish *Paramyxine sheni*. *Chromosome Res.*, 18, 383-400.

Kolkman, J. A. and Stemmer, W. P. C. (2001). Directed evolution of proteins by exon shuffling. *Nat. Biotech.*, 19, 423-428.

Kondo, T., Plaza, S., Zanet, J. et al. (2010). Small peptides switch the transcriptional activity of shavenbaby during *Drosophila* embryogenesis. *Science*, 329, 336-339.

Koopman, P., Munsterberg, A., Capel, B., Vivian, N. and Lovell-Badge, R. (1990). Expression of a candidate sex-determining gene during mouse testis differentiation. *Nature*, 348, 450-452.

Kordis, D. (2010). Transposable Elements in Reptilian and Avian (Sauropsida) Genomes. *Cytogenet. Genome Res.*, 127, 94-111.

Korenberg, J. R., Chen, X-N., Sun, Z., Shi, Z-Y., Ma, S., et al. (1999). Human genome anatomy: BACs integrating the genetic and cytogenetic maps for bridging genome and biomedicine. *Genome Res.*, 9, 994-1001.

Kouzarides, T. (2007). Chromatin modifications and their function. *Cell*, 128, 693-705.

Kuhn, G. C. S. and Sene, F. M. (2005). Evolutionary turnover of two pBuM satellite DNA subfamilies in the *Drosophila buzzatii* species cluster (repleta group): from alpha to alpha/beta arrays. *Gene*, 349, 77-85.

Kuhn, G. C.S., Sene, F. M., Moreira-Filho, O., Schwarzacher, T. and Heslop-Harrison, J. S. (2008). Sequence analysis, chromosomal distribution and long-range organization show that rapid turnover of new and old pBuM satellite DNA repeats leads to different patterns of variation in seven species of the *Drosophila buzzatii* cluster. *Chromosome Res.*, 16, 307-324.

Kuhn, G. C.S., Teo, C. H., Schwarzacher, T. and Heslop-Harrison, J. S. (2009). Evolutionary dynamics and sites of illegitimate recombination revealed in the interspersion and sequence junctions of two nonhomologous satellite DNAs in cactophilic *Drosophila* species. *Heredity*, 102, 453-464.

Kukekova, A., Vorobieva, N., Beklemisheva, V. et al. (2009). Chromosomal mapping of canine-derived BAC clones to the red fox and american mink genomes. *J. Hered.*, 42-53.

Kulemzina, A. I., Trifonov, V. A., Perelman, P. L. et al. (2009). Cross-species chromosome painting in Cetartiodactyla: Reconstructing the karyotype evolution in key phylogenetic lineages. *Chromosome Res.*, 17, 419-436.

Kunze, B., Traut, W., Garagna, S. et al. (1999). Pericentric satellite DNA and molecular phylogeny in Acomys (Rodentia). *Chromosome Res.*, 7, 131-141.

Kuramoto, M. (1990). A list of chromosome numbers of anurans amphibians. *Bull Fukuoka Univ. Educ. Nat. Sci.*, 39, 83-127.

Laan, M., Kallioniemi, P., Hellsten, E. et al. (1995). Mechanically stretched chromosomes as targets for high-resolution FISH mapping. *Genome Res.*, 5, 13-20.

Lander, E. S., Linton, L. M., Birren, B., and et al. (2001). Initial sequencing and analysis of the human genome. *Nature*, 409, 860-921.

Lanfredi, M., Congiu, L., Garrido-Ramos, M. A., et al. (2001). Chromosomal location and evolution of a satellite DNA family in seven sturgeon species. *Chromosome Res.*, 9, 47-52.

Lankenau, D-H. and Volff, J-N. (2009). Transposons and dynamic genome (184 p.). Berlin: Springer.

Lee, A. P., Koh, E. G., Tay, A., Brenner, S. and Venkatesh, B. (2006). Highly conserved syntenic blocks at the vertebrate Hox loci and conserved regulatory elements within and outside Hox gene clusters. *Proc. Natl. Acad. Sci. USA*, 103, 6994-6999.

Lee, B-Y. and Kocher, T. D. (2007). Exclusion of Wilms tumor (WT1b) and ovarian cytochrome P450 aromatase (CYP19A1) as candidates for sex determination genes in Nile tilapia (*Oreochromis niloticus*). *Anim. Genet.*, 38, 85-86.

Lee, B-Y., Penman, D. J. and Kocher, T. D. (2003). Identification of a sex-determining region in Nile tilapia (*Oreochromis niloticus*) using bulked segregant analysis. *Anim. Genet.*, 34, 379-383.

Legrand, J. J., Legrand-Hamelin, E. and Juchault, P. (1987). Sex determination in Crustacea. *Biol. Rev.*, 62, 439-470.

Levine, M. T., Jones, C. D., Kern, A. D., Lindfors, H. A. and Begun, D. J. (2006). Novel genes derived from noncoding DNA in *Drosophila melanogaster* are frequently X-linked and exhibit testis-biased expression. *Proc. Natl. Acad. Sci. USA*, 103, 9935-9939.

Li, R., Mignot, E., Faraco, J. et al. (1999). Construction and characterization of an eightfold redundant dog genomic bacterial artificial chromosome library. *Genomics,* 58, 9-17.

Li, W. H., Zhengllong, G., Wang, H. and Nekrutenko, A. (2001). Evolutionary analyses of the human genome. *Nature*, 409, 847-849.

Li, Y. C., Kord, A. B., Fahima, T., Berles, A. and Nero, E. (2002a). Microsatellites: genomic distribution, putative functions and mutation mechanisms: a review. *Mol. Ecol.*, 11, 2453-2465.

Li, Y. C., Lee, C., Chang, W. S., Li, S. Y. and Lin, C. C. (2002b). Isolation and identification of a novel satellite DNA family highly conserved in several Cervidae species. *Chromosoma,* 111, 176-183.

Li, Y. C., Lee, C., Sanoudou, D. and et al. (2000). Interstitial colocalization of two cervid satellite DNAs involved in the genesis in the Indian muntjac karyotype. *Chromosome Res.*, 8, 363-373.

Li, Y. X. and Kirby, M. L. (2003). Coordinated and conserved expression of alphoid repeat and alphoid repeat-tagged coding sequence. *Dev. Dyn.*, 228, 72-81.

Lichter, P., Cremer, T., Borden, J., Manuelidis, L. and Ward, D. C. (1988). Delineation of individual human chromosomes in metaphase and interphase cells by *in situ* suppression hybridization using recombinant DNA libraries. *Hum. Genet.*, 80, 224-234.

Lifton, R. P., Goldberg, M. L., Karp, R. W. and Hogness, D. S. (1978). The Organization of the Histone Genes in *Drosophila melanogaster:* Functional and Evolutionary Implications. *Quant. Biol.*, 42, 1047-1051.

Lin, C. C., Sasi, R., Fan, Y. S. and Chen, Z. Q. (1991). New evidence for tandem chromosome fusions in the karyotypic evolution of Asian muntjacs. *Chromosoma,* 101, 19-24.

Liu, Y., Nie, W. H., Huang, L. et al. (2008). Cloning, characterization, and FISH mapping of four satellite DNAs from Black muntjac (*Muntiacus crinifrons*) and Fea's muntjac (M. feae). *Zool. Res.*, 29, 225-235.

Locke, D. P., Archidiacono, N., Misceo, D. et al. (2003). Refinement of a chimpanzee pericentric inversion breakpoint to a segmental duplication cluster. *Genome Biol.*, 4, R50.

Long, E. O. and Dawid, I. B. (1980). Repeated genes in eukaryotes. *Ann. Rev. Biochem.*, 49, 727-764.

López-León, M. D., Cabrero, J., Dzyubenko, V. V. et al. (2008). Differences in ribosomal DNA distribution on A and B chromosomes between eastern and western populations of the grasshopper *Eyprepocnemis plorans plorans*. *Cytogenet. Genome Res.*, 121, 260-265.

López-León, M. D., Neves, N., Schwarzacher, T. et al. (1994). Possible origin of a B chromosome deduced from its DNA composition using double FISH technique. *Chromosome Res.*, 2, 87-92.

Loreto, V., Cabrero, J., López-León, M. D., Camacho, J. P. and Souza, M. J. (2008). Possible autosomal origin of macro B chromosomes in two grasshopper species. *Chromosome Res.*, 16, 233-241.

Lorite, P., Carrillo, J. A., Tinaut, A. and Palomeque, T. (2004). Evolutionary dynamics of satellite DNA in species of the Genus *Formica* (Hymenoptera, Formicidae). *Gene*, 332, 159-168.

Louzada, S., Paco, A., Kubickova, S. et al. (2008). Different evolutionary trails in the related genomes *Cricetus cricetus* and *Peromyscus eremicus* (Rodentia, Cricetidae) uncovered by orthologous satellite DNA repositioning. *Micron*, 39, 1149-1155.

Lucchini, S., Nardi, I., Barsacchi, G., Batistoni, R. and Andronico, F. (1993). Molecular cytogenetics of the ribosomal (18S + 28S and 5S) DNA loci in primitive and advanced urodele amphibians. *Genome*, 36, 762-773.

Luchetti, A., Cesar, M., Carrara, G. et al. (2003). Unisexuality and molecular drive: Bag320 sequence diversity in *Bacillus* taxa (Insecta Phasmatoidea). *J. Mol. Evol.*, 56, 587-596.

Lundin, L. G. (1999). Gene duplications in early metazoan evolution. *Cell Dev. Biol.*, 10, 523-530.

Lutz, B., Lu, H. C., Eichele, G., Miller, D. and Kaufman, T. C. (1996). Rescue of *Drosophila* labial null mutant by the chicken ortholog Hoxb-1 demonstrates that the function of Hox genes is phylogenetically conserved. *Genes Dev.*, 10, 176-184.

Lynch, J. M. (1971). Evolutionary relationships, osteology and zoogeography of leptodactylid frogs. *Univ. Kansas Mus. Nat. Hist. Misc. Publ.*, 53, 1-238.

Lynch, M. and Conery, J. S. (2003). The origins of genome complexity. *Science*, 302, 1401-1404.

Maggini, F. T., Cremonini, R., Zolfino, C. et al. (1991). Structure and chromosomal localization of DNA sequences related to ribosomal sub-repeats in *Vicia faba*. *Chromosoma*, 100, 229-234.

Malik, H. S., Burke, W. D. and Eickbush, T. H. (1999). The age and evolution of non-LTR retrotransposable elements. *Mol. Biol. Evol.*, 16, 793-805.

Mandrioli, M., Colomba, M. S. and Vitturi, R. (2000). Chromosomal analysis of repeated DNAs in the rainbow wrasse *Coris julis* (Pisces Labridae). *Genetica*, 108, 191-195.

Mank, J. E., Promislow, D. E. L. and Avise, J. C. (2006). Evolution of alternative sex-determining mechanisms in teleost fishes. *Biol. J. Linnean Soc.*, 87, 83-93.

Manolakou, P., Lavranos, G. and Angelopoulou, R. (2006). Molecular patterns of sex determination in the animal kingdom: a comparative study of the biology of reproduction. *Reprod. Biol. Endocrinol.*, 4, 1-23.

Marin, I., Plata-Rengifo, P., Labrador, M. and Fontdevila, A. (1998). Evolutionary relationships among the members of an ancient class of non-LTR retrotransposons found in the nematode *Caenorhabditis elegans*. *Mol. Biol. Evol.*, 15, 1390-1402.

Marshall, O. J., Chueh, C. A., Lee, H., Wong, L. H. and Choo, K. H. (2008). Neocentromeres: new insights into centromere structure, disease development, and karyotype evolution. *Am. J. Hum. Genet.*, 82, 261-282.

Martins, C. and Galetti Jr., P. M. (1999). Chromosomal localization of 5S rDNA genes in *Leporinus* fish (Anostomidae: Characiformes). *Chromosome Res.*, 7, 363-367.

Martins, C. and Galetti Jr., P. M. (2000). Conservative distribution of 5S rDNA loci in *Schizodon* (Pisces: Anostomidae) chromosomes. *Chromosome Res.*, 8, 353-355.

Martins, C. and Galetti, Jr., P. M. (2001). Organization of 5S rDNA in species of the fish *Leporinus*: two different genomic locations are characterized by distinct nontranscribed spacers. *Genome,* 44, 903-910.

Martins, C., Ferreira, I. A., Oliveira, C., Foresti, F. and Galetti Jr., P. M. (2006). A tandemly repetitive centromeric DNA sequence of the fish *Hoplias malabaricus* (Characiformes: Erythrinidae) is derived from 5S rDNA. *Genetica,* 127, 133-141.

Martins, C., Wasko, A. P., Oliveira, C. and Wright, J. M. (2000). Nucleotide sequence of 5S rDNA and localization of the ribosomal RNA genes to metaphase chromosomes of the tilapiine cichlid fish, *Oreochromis niloticus. Hereditas,* 133, 39-46.

Martins, C. and Wasko, A. P. (2004). Organization and evolution of 5S ribosomal DNA in the fish genome. In C. R. Williams (ed.), Focus on Genome Research (pp. 335-363). New York: Nova Science Publishers.

Mary-Elizabeth, M., Moy, G. W., Swanson, W. J. and Vacquier, V. D. (1998). Direct sequencing of genomic DNA for characterization of a satellite DNA in five species of Eastern Pacific abalone. *Mol. Mar. Biol. Biotechnol.,* 7, 1-6.

Maside, X., Bartolomé, C., Assimacopoulos, S. and CharlesWorth, B. (2001). Rates of movement and distribution of transposable elements in *Drosophila melanogaster*: in situ hybridization vs Southern blotting data. *Genet. Res.,* 78, 121-136.

Matsubara, K., Tarui, H., Toriba, M. et al. (2006). Evidence for different origin of sex chromosomes in snakes, birds, and mammals and step-wise differentiation of snake sex chromosomes. *Proc. Natl. Acad. Sci. USA,* 103, 18190-18195.

Matsubara, K., Yamada, K., Umemoto, S. et al. (2008). Molecular cloning and characterization of the repetitive DNA sequences that comprise the constitutive heterochromatin of the A and B chromosomes of the Korean field mouse (*Apodemus peninsulae*, Muridae, Rodentia). *Chromosome Res.,* 16, 1013-1026.

Matsuda, M. (2005). Sex determination in the teleost medaka, *Oryzias latipes. Ann. Rev. Genet.,* 39, 293-307.

Matsuda, M., Nagahama, Y., Shinomiya, A. et al. (2002). DMY is a Y-specific DM-domain gene required for male development in the medaka fish. *Nature,* 417, 559-563.

Matsuda, Y., Nishida-Umehara, C., Tarui, H., et al. (2005). Highly conserved linkage homology between birds and turtles: bird and turtle chromosomes are precise counterparts of each other. *Chromosome Res.,* 13, 601-615.

Matsui, M., Seto, T., Kohsaka, Y. and Borkin, L. J. (1995). Bearing of chromosome C-banding patterns on the classification of Eurasian toads ofthe *Bufo bufo* complex. *Amph. Rept.,* 16, 23-33.

Matthey, R. (1973). The chromosome formula of eutherian mammls. In Chiarelli, A. B. and Capanna, E. (eds.), Cytotaxonomy and vertebrate evolution (pp. 531-616). London: Academic Press.

Matzke, A. J. M., Varga, F., Gruendler, P. et al. (1992). Characterization of a new repetitive sequence that is enriched on microchromosomes of turkey. *Chromosoma,* 102, 9-14.

Matzke, M. A., Varga, F., Berger, H. et al. (1990). A 41-42 bp tandemly repeated sequence isolated from nuclear envelopes of chicken erythrocytes is located predominantly on microchromosomes. *Chromosoma,* 99, 131-137.

Maxson, R., Cohn, R. and Kedes, L. (1983). Expression and organization of histone genes. *Ann. Rev. Genet.,* 17, 239-277.

Mazzuchelli, J. and Martins, C. (2009). Genomic organization of repetitive DNAs in the cichlid fish *Astronotus ocellatus*. *Genetica,* 136, 461-469.

McAllister, B. F. and Werren, J. H. (1997). Hybrid origin of a B chromosome (PSR) in the parasitic wasp *Nasonia vitripennis*. *Chromosoma,* 106, 243-253.

McClung, C. E. (1901). Notes on the accessory chromosome. *Anat. Anz.,* 20, 220-226.

McClung, C. E. (1902). The accessory chromosome-sex determinant? *Biol. Bull,* 3, 43-84.

McDaniell, R., Lee, B. K., Song, L. et al. (2010). Heritable Individual-Specific and Allele-Specific Chromatin Signatures in Humans. *Science,* 328, 235-239.

McFadyen, D. A. and Locke, J. (2000). High-resolution FISH mapping of the rat α2u-globulin multigene family. *Mamm. Genome,* 11, 292-299.

McMillan, D., Miethke, P., Alsop, A. E. et al. (2007). Characterizing the chromosomes of the platypus (*Ornithorhynchus anatinus*). *Chromosome Res.,* 15, 961-974.

McNeil, N. and Ried, T. (2000). Novel molecular cytogenetic techniques for identifying complex chromosomal rearrangements: technology and applications in molecular medicine. *Expert Rev. Mol. Med.,* 2, 1-14.

McQuade, L., Hill, R. and Francis, D. (1994). B chromosome systems in the greater glider, *Petauroides volans* (Marsupialia, Pseudocheiridae). 2. Investigation of B chromosome DNA sequences isolated by micromanipulation and PCR. *Cytogenet. Cell Genet.,* 66, 155-161.

Melaragno, J. E., Mehrotra, B. and Coleman, A. W. (1993). Relationship between endopolyploidy and cell size in epidermal tissue of *Arabidopsis*. *Plant Cell,* 5, 1661-1668.

Meltzer, P. S., Guan, X. Y., Burgess, A. and Trent, J. M. (1992). Rapid generation of region specific probes by chromosome microdissection and their application. *Nat. Genet.,* 1, 24-28.

Mesa, A., Ferreira, A. and Carbonell, C. S. (1982). Cariología de los acridoideos neotropicales: estado actual de su conocimiento y nuevas contribuciones. *Annls. Soc. Ent. Fr.,* 18, 507-526.

Mestriner, C. A., Galetti Jr., P. M., Valentini, S. R. et al. (2000). Structural and functional evidence that a B chromosome in the characid fish *Astyanax scabripinnis* is an isochromosome. *Heredity,* 85, 1-9.

Meyne, J., Baker, R. J., Hobart, H. H. et al. (1990). Distribution of non-telomeric sites of the (TTAGGG)n telomeric sequence in vertebrate chromosomes. *Chromosoma,* 99, 3-10.

Michailova, P., Ilkova, J., Hankeln, T. et al. (2009). Somatic breakpoints, distribution of repetitive DNA and non-LTR retrotransposon insertion sites in the chromosomes of *Chironomus piger* Strenzke (Diptera, Chironomidae). *Genetica,* 135, 137-148.

Millar, C. D., Lambert, D. M., Bellamy, A. R., Stapleton, P. M. and Young, E. C. (1992). Sex-specific restriction fragments and sex ratios revealed by DNA fingerprinting In the Brown skua. *J. Hered.,* 83, 350-355.

Miller, D. J., Harrison, P. L., Mahony, T. J. et al. (1993). Nucleotide sequence of the histone gene cluster in the coral *Acropora formosa* (Cnidaria; Scleractinia): features of histone gene structure and organization are common to diploblastic and triploblastic metazoans. *J. Mol. Evol.* 37, 245-253.

Miller, K., Lynch, C., Martin, J., Herniou, E. and Tristem, M. (1999). Identification of multiple *gypsy* LTRretrotransposon lineages in vertebrate genomes. *J. Mol. Evol.,* 49, 358-366.

Ming, R. and Moore, P. H. (2007). Genomics of sex chromosomes. *Curr. Opin. Plant Biol.,* 10, 123-130.

Mingazzini, V., Luchetti, A. and Mantovani, B. (2010). R2 dynamics in Triops cancriformis (Bosc, 1801) (Crustacea, Branchiopoda, Notostraca): turnover rate and 28S concerted evolution. *Heredity* (*in press*).

Mitra, A., Liu, G. and Song, J. (2009). A genome-wide analysis of array-based comparative genomic hybridization (CGH) data to detect intra-species variations and evolutionary relationships. *PLoS One*, 4, e7978.

Miura, I. (1995). Two differentiated groups of the Japanese toad, *Bufo japonicus japonicus*, demonstrated by C-banding analysis of chromosomes. *Caryologia*, 2, 123-136.

Miyake, T. and Amemiya, C. T. (2004). BAC libraries and comparative genomics of aquatic chordate species. *Comp. Biochem. Physiol.*, 138, 233-244.

Molina, W. F. (2006). Chromosomal changes and stasis in marine fish groups. In E. Pisano, C. Ozouf-Costaz, F. Foresti and B. G. Kapoor (eds.), Fish cytogenetics (pp. 69-110). Enfield, New Hampshire: Science Publisher.

Morescalchi, A. (1968). The karyotype of two specimes of *Leiopelma hochstetteri* Fitz. (Amphibia, Salientia). *Caryologia*, 21, 37-46.

Morescalchi, A. (1973) Amphibia. In A. B. Chiarelli and E. Capanna (eds.), Cytotaxonomy and vertebrate evolution (pp. 233-248). London, New York: Academic Press.

Morescalchi, A. (1975). Chromosome evolution in the caudate Amphibia. *Evol. Biol.*, 8, 339-387.

Morgan, T. H., Sturtevant, A. H., Bridges, C. B. and Muller, H. J. (1915). The mechanism of mendelian heredity. New York, NY: Holt Rinehart and Winston. Reprinted. Johnson Reprint Corporation with an Introduction by Garland E. Allen, 1978.

Mouse Genome Sequencing Consortium (2002). *Nature*, 420, 520.

Mravinac, B., Plohl, M. and Ugarković, D. (2004). Conserved patterns in the evolution of *Tribolium* satellite DNAs. *Gene*, 332, 169-177.

Mravinac, B., Ugarković, D., Franjević, D. and Plohl, M. (2005). Long inversely-oriented subunits form a complex monomer of *Tribolium brevicornis* satellite DNA. *J. Mol. Evol.*, 60, 513-525.

Mueller, U. G. and Wolfenbarger L. W. (1999) AFLP genotyping and fingerprinting. *Tree*, 14, 389-394.

Muller, H. J. (1940). Bearings of the '*Drosophila*' work on systematics. In J. Huxley (ed.), The New Systematics (pp. 185-268). Oxfor: Clarendon Press.

Muller, S., Stanyon, R., O'Brien, P. C., Ferguson-Smith, M. A., Plesker, R. and Wienberg, J. (1999). Defining the ancestral karyotype of all primates by multidirectional chromosome painting between tree shrews, lemurs and humans. *Chromosoma*, 108, 393-400.

Murphy, W. J., Stanyon, R. and O'Brien, S. J. (2001). Evolution of mammalian genome organization inferred from comparative gene mapping. *Genome Biol.*, 2, 6.

Myers, P. (2008). Synteny: Inferring ancestral genomes. *Nat. Educ.*, 1, 1.

Naciri, Y., Vigouroux, Y., Dallas, J. et al. (1995). Identification and inheritance of (GA/TC)n and (AC/GT)n repeats in the European flat oyster *Ostrea edulis* (L.). *Mol. Mar. Biol. Biotechnol.*, 4, 83-89.

Nagoda, N., Fukuda, A., Nakashima, Y. and Matsuo, Y. (2005). Molecular characterization and evolution of the repeating units of histone genes in *Drosophila americana*: coexistence of quartet and quintet units in a genome. *Insect Mol. Biol*, 14, 6, 713-717.

Nakai, Y., Kubota, S., Goto, Y. et al. (1995). Chromosome elimination in three Baltic, south Pacific and north-east Pacific hagfish species. *Chromosome Res.*, 3, 321-330.

Nakamura, M. (2009). Sex determination in amphibians. *Sem. Cell Develop. Biol.*, 20, 271-282.

Nakayama, I., Foresti, F., Tewari, R., Schartl, M. and Chourrout, D. (1994). Sex chromosome polymorphism and heterogametic males revealed by two cloned DNA probes in the ZW/ZZ fish *Leporinus elongatus*. *Chromosoma*, 103, 31-39.

Nanda, I., Feichtinger, W., Schmid, M. et al. (1990). Simple repetitive sequences are associated with the differentiation of the sex chromosomes in the guppy fish. *J. Mol. Evol.*, 30, 456-462.

Nanda, I., Karl, E., Griffin, D. K., Schartl, M. and Schmid, M. (2007). Chromosome repatterning in three representative parrots (Psittaciformes) inferred from comparative chromosome painting. *Cytogenet. Genome Res.*, 117, 43-53.

Nanda, I., Karl, E., Volobouev, V. et al. (2006). Extensive gross genomic rearrangements between chicken and Old World vultures (Falconiformes: Accipitridae). *Cytogenet. Genome Res.*, 112, 286-295.

Nanda, I., Schlegelmilch, K., Haaf, T., Schartl, M. and Schmid, M. (2008). Synteny conservation of the Z chromosome in 14 avian species (11 families) supports a role for Z dosage in avian sex determination. *Cytogenet. Genome Res.*, 122, 150-156.

Nanda, I., Shan, Z., Schartl, M. et al. (1999). 300 million years of conserved synteny between chicken Z and human chromosome 9. *Nat. Genet.*, 21, 258-259.

Nanda, I., Zend-Ajusch, E., Shan, Z. et al. (2000). Conserved synteny between the chicken Z sex chromosome and human chromosome 9 includes the male regulatory gene DMRT1: a comparative (re)view on avian sex determination. *Cytogenet. Cell Genet.*, 89, 67-78.

Nash, W. G., Wienberg, J., Ferguson-Smith, M. A., Menninger, J. C. and O'Brien, S. J. (1998). Comparative genomics: tracking chromosome evolution in the family Ursidae using reciprocal chromosome painting. *Cytogenet. Cell Genet.*, 83, 182-192.

Nederlof, P. M., Robinson, D., Abuknesha, R. et al. (1989). Three-color fluorescence in situ hybridization for the simultaneous detection of multiple nucleic acid sequences. *Cytometry*, 10, 20-27.

Neitzel, H., Kalscheuer, V., Henschel, S., Digweed, M. and Sperling, K. (1998). Beta-heterochromatin in mammals: evidence from studies in *Microtus agrestis* based on the extensive accumulation of L1 and non-L1 retroposons in the heterochromatin. *Cytogenet. Genome Res.*, 80, 165-172.

Neitzel, H., Kalscheuer, V., Singh, A. P., Henschel, S. and Sperling, K. (2002). Copy and paste: the impact of a new non-L1 retroposon on the gonosomal heterochromatin of *Microtus agrestis*. *Cytogenet. Genome Res.*, 96, 179-185.

Neuhaus, H., Müller, F., Etter, A. and Tobler, H. (1987). Type I-like intervening sequences are found in the rDNA of the nematode *Ascaris lumbricoides*. *Nucleic Acids Res.*, 15, 7689-7707.

Nguyen, P., Sahara, K., Yoshido, A. and Marec, F. (2010). Evolutionary dynamics of rDNA clusters on chromosomes of moths and butterflies (Lepidoptera). *Genetica*, 138, 343-354.

Nie, W., Wang, J., O'Brien, P. C. M. et al. (2002). The genome phylogeny of domestic cat, red panda and five mustelid species revealed by comparative chromosome painting and G-banding. *Chromosome Res.*, 10, 209-222.

Nishida-Umehara, C., Tsuda, Y., Ishijima, J. et al. (2007). The molecular basis of chromosome orthologies and sex chromosomal differentiation in palaeognathous birds. *Chromosome Res.*, 15, 721-734.

Nishioka, M. and Hanada, H. (1994). Sex of reciprocal hybrids between the Hamakita (XX-XY type) population and the Murakami (ZW-ZZ type) population of *Rana rugosa. Sci. Rep. Lab. Amphib. Biol. Hiroshima. Univ.,* 13, 35-50.

Nishioka, M., Hanada, H., Miura, I. and Ruyzaki, M. (1994). Four kinds of sex chromosomes in *Rana rugosa. Sci. Rep. Lab. Amphib. Biol. Hiroshima Univ.,* 13, 1-34.

Nokkala, S., Grozeva, S., Kuznetsova, V. and Maryanska-Nadachowska, A. (2003). The origin of the achiasmatic XY sex chromosome system in *Cacopsylla peregrina* (Frst.) (Psylloidea, Homoptera). *Genetica,* 119, 327-332.

Nokkala, S., Kuznetsova, V. and Maryanska-Nadachowska, A. (2000). Achiasmate segregation of a B chromosome from the X chromosome in two species of psyllids (Psylloidea: Homoptera). *Genetica,* 108, 181-189.

Nomoto, Y., Hirai, M. and Ueshima, R. (2001). Cloning of Molluscan telomere DNA with $(TTAGGG)_n$ repeat and its chromosomal location in the freshwater snail *Biwamelania habei. Zool. Sci.,* 18, 417-422.

Novick, P. A., Basta, H., Floumanhaft, M., McClure, M. A. and Boissinot, S. (2009). The evolutionary dynamics of autonomous non-LTR retrotransposons in the lizard *Anolis carolinensis* shows more similarity to fish than mammals. *Mol. Biol. Evol.*, 26, 1811-1822.

Nowak, R. (1994). Mining treasures from junk DNA. *Science*, 263, 608-610.

O'Brien, S. J. and Stanyon, R. (1999). Phylogenomics. Ancestral primate viewed. *Nature,* 402, 365-366.

O'Brien, S. J., Nash, W. G. and Menninger, J. C. (2006). Atlas of mammalian chromosomes (714 p.). Hoboken: Wiley-Liss.

O'Meally, D., Patel, H. R., Stiglec, R. et al. (2010). Non-homologous sex chromosomes of birds and snakes share retitive sequences. *Chromosome Res., in press.*

Ocalewicz, K., Mota-Velasco, J. C., Campos-Ramos, R. and Penman, D. J. (2009). FISH and DAPI staining of the synaptonemal complex of the Nile tilapia (*Oreochromis niloticus*) allow orientation of the unpaired region of bivalent 1 observed during early pachytene. *Chromosome Res.*, 17, 773-782.

Odierna, G., Aprea, G., Capriglione, T. and Balletto, E. (2004). Evidence for chromosome and *Pst*I satellite DNA family evolutionary stasis in the *Bufo viridis* group (Amphibia, Anura). *Chromosome Res.*, 12, 671-681.

Oguiura, N., Ferrarezzi, H. and Batistic, R. F. (2010). Cytogenetics and molecular data in snakes: A phylogenetic approach. *Cytogenet. Genome Res.*, 127, 128-142.

Ohno, S. (1967). Sex chromosomes and sex linked genes (192 p.). Berlin: Springer Verlag.

Ohno, S. (1970). Evolution by gene duplication (160 p.). New York: Springer.

Ohtani, H. (1993). Mechanism of chromosome elimination in the hybridogenetic spermatogenesis of allotriploid males between Japanese and European water frogs. *Chromosoma*, 102, 158-162.

Okazaki, S., Tsuchida, K., Maekama, H., Ishikawa, H. and Fujiwara, H. (1993). Identification of a Pentanucleotide Telomeric Sequence, $(TTAGG)n$, in the Silkworm *Bombyx mori* and in Other Insects. *Mol. Cel. Biol.* 12, 1424-1432.

Oliveira, C., Almeida-Toledo, L. F. and Foresti, F. (2007). Karyotypic evolution in Neotropical fishes. In: E. Pisano, C. Ozouf-Costaz, F. Foresti and B. G. Kapoor (eds.), Fish cytogenetics (pp. 111-164). Enfield, New Hampshire: Science Publisher.

Olmo, E. (1986). Chordata 3. A. Reptilia. Animal Cytogenetics, vol. 4. (100 p.). Berlin: Gebrüder Borntraeger.

Olmo, E. and Signorino, G. (2005). Chromorep: a reptile chromosomes database. 2010-10-30. Available from: http://193.206.118.100/professori/chromorep.pdf

Olmo, E., Capriglione, T. and Odierna, G. (2002). Different genomic evolutionary rates in the various reptile lineages. *Gene*, 295, 317-321.

Ono, T., Kondoh, Y., Kagiyama, N., Sonta, S. and Yoshida, M. C. (2001). Genomic organization and chromosomal distribution of rat ID elements. *Genes Genet. Syst.*, 76, 213-220.

Ozouf-Costaz, C., Brandt, J., Körting, C. et al. (2004). Genome dynamics and chromosomal localization of the non-LTR retrotransposons Rex1 and Rex3 in Antarctic fish. *Antarct. Sci.*, 16, 51-57.

Palomeque, T. and Lorite, P. (2008). Satellite DNA in insects: a review. *Heredity*, 100, 564-573.

Panicker, S. G. and Singh, L. (1994). Banded krait minor satellite (Bkm) contains sex and species-specific repetitive DNA. *Chromosoma*, 103, 40-45.

Pannebakker, A., Niehuis, O., Hedley, A., Gadau, J., and Shuker, D. M. (2010). The distribution of microsatellites in the *Nasonia* parasitoid wasp genome. *Insect Mol. Biol.*, 19, 91-98.

Papeschi, A. G. and Bressa, M. J. (2006). Evolutionary cytogenetics in Heteroptera. *J. Biol. Res.*, 5, 3-21.

Pardini, A. T., O'Brien, P. C. M., Fu, B. et al. (2007). Chromosome painting among Proboscidea, Hyracoidea and Sirenia: support for Paenungulata (Afrotheria, Mammalia) but not Tethytheria. *Proc. R. Soc. B*, 274, 1333-1340.

Pardue, M. L. and Gall, J. G. (1969). Molecular hybridization of radioactive DNA to the DNA of cytological preparations. *Proc. Natl. Acad. Sci. USA*, 64, 600-604.

Patel, N. H. and Prince, V. E. (2000). Beyond the Hox complex. *Genome Biol*, 1, 1027.1-1027.4.

Patthy, L. (1999). Genome evolution and the evolution of exon-shuffling — a review. *Gene*, 238, 103-114.

Pauls, E. and Bertollo, L. A. (1983). Evidence for a system of supernumerary chromosomes in *Prochilotus scrofa* Steindachner, 1881 (Pisces, Prochilodontidae). *Caryologia*, 36, 307-314.

Pedersen, C. and Linde-Laursen, I. (1995). The relationship between physical and genetic distances at the Hor1 and Hor2 loci of barley estimated by two-color fluorescent in situ hybridization. *Theor. Appl. Genet.*, 91, 941-946.

Pedrosa, A., Vallejos, C. E., Bachmair, A. and Schweizer, D. (2003). Integration of common bean (*Phaseolus vulgaris* L.) linkage and chromosomal maps. *Theor. Appl. Genet.*, 106, 205-12.

Pellestor, F., Girardet, A., Lefort, G., Andréo, B. and Charlieu, J. P. (1995). Use of the primed in situ labeling (PRINS) technique for a rapid detection of chromosomes 13, 16, 18, 21, X and Y. *Hum. Genet.*, 95, 12-17.

Pellestor, F., Girardet, A., Andréo, B. and Charlieu, J. P. (1994). Apolimorphic alfa-satellite sequence for human chromosome 13 detected by oligonucleotide primed in situ labeling (PRINS). *Hum. Genet.*, 94, 346-348.

Pendás, A. M., Morán, P. and García-Vázquez, E. (1994). Organization and chromosomal location of the major histone cluster in brown trout, Atlantic salmon and rainbow trout. *Chromosoma*, 103, 147-152.

Peppers, J. A., Wiggins, L. E. and Baker, R. J. (1997). Nature of B chromosomes in the harvest mouse *Reithrodontomys megalotis* by fluorescence *in situ* hybridization (FISH). *Chromosome Res.*, 5, 475-479.

Perani, P., Caccio, S., Saccone, S., Andreozzi, L. and Bernardi, G. (2000). Telomeres in warm-blooded vertebrates are composed of GC-rich isochores. *Biochem. Genet.*, 38, 227-239.

Perelman, P. L., Graphodatsky, A. S., Dragoo, J. W. et al. (2008). Chromosome painting shows that skunks (Mephitidae, Carnivora) have highly rearranged karyotypes. *Chromosome Res.*, 16, 215-1231.

Perelman, P. L., Graphodatsky, A. S., Serdukova, N. A. et al. (2005). Karyotypic conservatism in the suborder Feliformia (Order Carnivora). *Cytogenet. Genome Res.*, 108, 348-354.

Pérez, J., Morán, P. and García-Vasquez, E. (1999). Physical mapping of three minisatellite sequences in the Atlantic salmon (*Salmo salar*) genome. *Anim. Genet.*, 30, 371-374.

Perfectti, F. and Werren, J. H. (2001). The interspecific origin of B chromosomes: experimental evidence. *Evolution*, 55, 1069-1073.

Peterson, B. K., Hare, E. E., Iyer, V. N. et al. (2009). Big genomes facilitate the comparative identification of regulatory elements. *PLoS One*, 4, e4688.

Peterson, E. T., Sutherland, R., Robinson, D. L. et al. (1999). An integrated physical map for the short arm of human chromosome 5. *Genome Res.*, 9, 1250-1267.

Petrov, D. A. (2001). Evolution of genome size: new approaches to an old problem. Opinion, *Trends Genet.*, 17, 23-28.

Petrov, D. A., Sangster, T. A., Johnston, J. S., Hartl, D. L. and Shaw, K. L. (2000). Evidence for DNA loss as a determinant of genome size. *Science*, 287, 1060-1062.

Petrović, V., Pérez-García, C., Pasantes, J. J. et al. (2009). A GC-rich satellite DNA and karyology of the bivalve mollusk *Donax trunculus*: a dominance of GC-rich heterochromatin. *Cytogenet. Genome Res.*, 124, 63-71.

Pevzner, P. and Tesler, G. (2003). Genome rearrangements in mammalian evolution: Lessons from human and mouse genomes. *Genome Res.*, 13, 37-45.

Phillips, R. B. (2001). Application of fluorescence in situ hybridization (FISH) to fish genetics and genome mapping. *Mar. Biotechnol.*, 3, 145-152.

Phillips, R. B. and Reed, K. M. (1996). Application of fluorescence *in situ* hybridization (FISH) techniques to fish genetics: a review. *Aquaculture*, 140, 197-216.

Phillips, R. B., Keatley, K. A., Morasch, M. R. et al. (2009). Assignment of Atlantic salmon (*Salmo salar*) linkage groups to specific chromosomes: Conservation of large syntenic blocks corresponding to whole chromosome arms in rainbow trout (*Oncorhynchus mykiss*). *BMC Genet.*, 10, 46.

Phillips, R. B., Pleyte, K. A. and Hartley, S. E. (1988). Stock-specific differences in the number and chromosome positions of the nucleolar organizer regions in Arctic charr (*Salvelinus alpinus*). *Cytogenet. Cell Genet.*, 48, 9-12.

Picariello, O., Feliciello, I., Bellinero, R. and Chinali, G. (2002). S1 satellite DNA as a taxonomic marker in brown frogs: molecular evidence that *Rana graeca graeca* and *Rana graeca italica* are diferent species. *Genome*, 45, 63-70.

Pigozzi, M. I. (2007). Localization of single-copy sequences on chicken synaptonemal complex spreads using fluorescence *in situ* hybridization (FISH). *Cytogenet. Genome Res.*, 119, 105-112.

Pigozzi, M. I. (2008). Relationship between physical and genetic distances along the zebra finch Z chromosome. *Chromosome Res.*, 16, 839-849.

Pimpinelli, S., Berlocot, M., Fantit, I. et al. (1995). Transposable elements are stable structural components of *Drosophila melanogaster* heterochromatin. *Proc. Natl. Acad. Sci. USA*, 92, 3804-3808.

Pineau, P., Henry, M., Suspène, R. et al. (2005). A universal primer set for PCR amplification of nuclear histone H4 genes from all animal species. *Mol. Biol. Evol.*, 22, 582-588.

Pinkel, D., Landegent, J., Collins, C. et al. (1988). Fluorescence in situ hybridization with human chromosome-specific libraries detection of trisomy-21 and translocations of chromosome-4. *Proc. Nat. Acad. Sci. USA*, 85, 9138-9142.

Pinkel, D., Straume, T. and Gray, J. W. (1986). Cytogenetic analysis using quantitative, high-sensitivity, fluorescence hybridization. *Proc. Natl. Acad. Sci. USA*, 83, 2934-2938.

Pinton, A., Ducos, A., Séguéla, A. et al. (1998). Characterization of reciprocal translocations in pigs using dual-color chromosome painting and primed in situ DNA labelling. *Chromosome Res.*, 6, 361-366.

Piras, F. M., Nergadze, S. G., Magnani, E. et al. (2010). Uncoupling of Satellite DNA and Centromeric Function in the Genus *Equus*. *PLoS Genetics*, 6, e1000845.

Pisano, E., and Ghigliotti, L. (2009). Ribosomal genes in notothenioid fishes: Focus on the chromosomal organization. *Mar. Genomics*, 2, 75-80.

Pisano, E., Cocca, E., Mazzei, F. et al. (2003). Mapping of α- and β-globin genes on Antarctic fish chromosomes by fluorescence *in-situ* hybridization. *Chromosome Res.*, 11, 633-640.

Piskurek, O., Nishihara, H. and Okada, N. (2009). The evolution of two partner LINE/SINE families and a full-length chromodomain-containing Ty3/ *Gypsy* LTR element in the first reptilian genome of *Anolis carolinensis*. *Gene*, 441, 111-118.

Plohl, M. and Cornudella, L. (1997). Characterization ofinter-related sequence motifs in four satellite DNAs and their distribution in the genome of the mollusk *Donax trunculus*. *J. Mol. Evol.*, 44, 189-198.

Plohl, M., Luchetti, A., Mestrovic, N. and Mantovani, B. (2008). Satellite DNAs between selfishness and functionality: structure, genomics and evolution of tandem repeats in centromeric (hetero) chromatin. *Gene*, 409, 72-82.

Poletto, A. B., Ferreira, I. A., Cabral-de-Mello, D. C. et al. (2010a). Chromosome differentiation patterns during cichlid fish evolution. *BMC Genet.*, 11, 50.

Poletto, A. B., Ferreira, I. A. and Martins, C. (2010b). The B chromosomes of the African cichlid fish *Haplochromis obliquidens* harbour 18S rRNA gene copies. *BMC Genet.*, 11, 1.

Pontecorvo, G., De Felice, B. and Carfagna, M. (2000). A novel repeated sequence DNA originated from a Tc1-like transposon in water green frog Rana esculenta. *Gene*, 261, 205-210.

Ponzi, M., Pace, T., Dore, E. and Frontau, C. (1985). Identification of telomeric DNA sequence in *Plasmodium berghei*. *EMBO J.* 4, 2991-2995.

Poptsova, M. S. and Gogarten, J. P. (2007). The power of phylogenetic approaches to detect horizontally transferred genes. *BMC Evol. Biol.*, 7, 1-17.

Postlethwait, J. H., Yan, Y., Gates, M. A., Horne, S. et al. (1998). Vertebrate genome evolution and the zebrafish gene map. *Nat. Genet.*, 18, 345-349.

Powers, T. P. and Amemiya, C. T. (2004). Evidence for Hox 14 paralog group in vertebrates. *Curr. Biol.*, 14, 183-184.

Primmer, C., Raudsepp, T., Chowdhary, B. P., Møller, A. P. and Ellegren, H. (1997). Low Frequency of Microsatellites in the Avian Genome. *Genome Res.*, 7, 471-482.

Pritham, E. J. (2009). Transposable Elements and Factors Influencing their Success in Eukaryotes. *J. Hered.*, 100, 648-655.

Prokopowich, C. D., Gregory, T. R. and Crease, T. J. (2003). The correlation between rDNA copy number and genome size in eukaryotes. *Genome*, 46, 48-50.

Putnam, N. H., Srivastava, M., Hellsten, U. and et al. (2007). Sea anemone genome reveals ancestral eumetazoan gene repertoire and genomic organization. *Science*, 317, 86-94.

Quiniou, S. M. A., Katagiri, T., Miller, N. W. et al. (2003). Construction and characterization of a BAC library from a gynogenetic channel catfish, *Ictalurus punctatus*. *Genet. Sel. Evol.*, 35, 1-11.

Qumsiyeh, M. B. (1994). Evolution of number and morphology of mammalian chromosomes. *J. Hered*, 85, 455-465.

Qureshi, S. A. and Blake, R. D. (1995). Sequence characteristics of a cervid DNA repeat family. *J. Mol. Evol.*, 40, 400-404.

Rabbitts, P., Impey, H., Heppell-Parton, A. et al. (1995). Chromosome specific paints from a high resolution flow karyotype of the mouse. *Nat. Genet.*, 9, 369-375.

Rabenold, P. P., Rabenold, K.N., Piper, W. H., Haydock, J. and Zack, S.W. (1990). Shared paternity revealed by genetic analysis in cooperatively breeding tropical wrens. *Nature*, 348, 538-540.

Rabl, C. (1885). Uber Zelltheilung. *Morphologisches Jahrbuch,* 10, 214-330.

Rajcan-Separovic, E. and Sabour, M. P. (1993). Fluorescence *in situ* hybridization of bovine *Alu*-like sequences to bovine and ovine chromosomes. *Genome*, 36, 984-986.

Ranz, J. M., González, J., Casals, F. and Ruiz, A. (2003). Low occurrence of gene transposition events during the evolution of the genus *Drosophila*. *Evolution*, 57, 1325-1335.

Raskina, O., Barber, J. C., Nevo, E. and Belyayev, A. (2008). Repetitive DNA and chromosomal rearrangements: speciation-related events in plant genomes. *Cytogenet. Genome Res.*, 120, 351-357.

Rebollo, E., Martin, S., Manzanero, S. and Arana, P. (1998). Chromosomal strategies for adaptation to univalency. *Chromosome Res.,* 6, 515-531.

Reece, R. J. (2004). Analysis of Genes and Genomes (469 p.). Chichester: John Wiley & Sons.

Rens, W., Grützner, F., O'Brien, P. C. et al. (2004). Resolution and evolution of the duck-billed platypus karyotype with an $X_1Y_1X_2Y_2X_3Y_3X_4Y_4X_5Y_5$ male sex chromosome constitution. *Proc. Natl. Acad. Sci. USA*, 101, 16257-16261.

Rens, W., O'Brien, P. C., Yang, F. et al. (2001). Karyotype relationships between distantly related marsupials from South America and Australia. *Chromosome Res.*, 9, 301-308.

Rens, W., O'Brien, P. C. M., Grützner, F. et al. (2007). The multiple sex chromosomes of platypus and echidna are not completely identical and several share homology with the avian Z. *Genome Biol.*, 8, R243.

Richard, F., Lombard, M. and Dutrillaux, B. (2003). Reconstruction of the ancestral karyotype of eutherian mammals. *Chromosome Res.*, 11, 605-618.

Richard, G-F., Kerrest, A. and Dujon, B. (2008). Comparative genomics and molecular dynamics of DNA repeats in eukaryotes. *Microbiol. Mol. Biol. Rev.*, 72, 686-727.

Richards, E. J. and Ausubel, F. M. (1988). Isolation of a higher eukaryotic telomere from *Arabidopsis thalliana*. *Cell*, 53, 127-136.

Ridley, M. (1996). Evolution (2nd ed. 458 p.). Oxford: Oxford University Press.

Robinson, T. J. and Ruiz-Herrera, A. (2008). Defining the ancestral eutherian karyotype: A cladistic interpretation of chromosome painting and genome sequence assembly data. *Chromosome Res.*, 16, 1133-1141.

Robinson, T. J., Fu, B., Ferguson-Smith, M. A. and Yang, F. (2004). Cross-species chromosome painting in the golden mole and elephant shrew: support for the mammalian clades Afrotheria and Afroinsectiphilla but not Afroinsectivora. *Proc. R. Soc. Lond. B Biol. Sci.*, 271, 1477-1484.

Röder, M., Korzun, V., Wendehake, K. et al. (1998). A microsatellite map of wheat. *Genetics*, 149, 2007-2023.

Rodionov, A. V. (1997). Evolution of avian chromosomes and linkage groups. *Russ. J. Genet.*, 33, 605-617.

Roiha, H., Miller, J. R., Woods, L. C. and Glover, D. M. (1981). Arrangements and rearrangements of sequences flanking the two types of rDNA insertion in *D. melanogaster*. *Nature*, 290, 749-753.

Romanov, M. N., Daniels, L. M., Dodgson, J. B. and Delany, M. E. (2005). Integration of the cytogenetic and physical maps of chicken chromosome 17. *Chromosome Res.*, 13, 215-222.

Ropiquet, A., Gerbault-Seureau, M., Deuve, J. L. et al. (2008). Chromosome evolution in the subtribe Bovina (Mammalia, Bovidae): The karyotype of the Cambodian banteng (*Bos javanicus birmanicus*) suggests that Robertsonian translocations are related to interspecific hybridization. *Chromosome Res.*, 16, 1107-1118.

Rozen, S. and Skaletsky, H. (2000). Primer3 on the WWW for general users and for biologist programmers. *Methods Mol. Biol.*, 132, 365-86.

Ruberti, I., Fragapane, P., Pierandrei-Amaldi, P. et al. (1982). Characterization of histone genes isolated from *Xenopus laevis* and *Xenopus tropicalis* genomic libraries. *Nucleic Acids Res.*, 10, 1544-1550.

Rudykh, I. A., Grechko, V. V., Kramerov, D. A. and Darevsky, I. S. (1999). Distribution of *Hind*III-repeats in genomes of Caucasian lizards of the *Lacerta* species reflects their phylogenetic affiliation. *Dokl. Akad. Nauk*, 367, 563-566.

Ruiz-Herrera, A., Castresana, J. and Robinson, T. J. (2006). Is mammalian chromosomal evolution driven by regions of genome fragility? *Genome Biol.*, 7, R115.

Ruiz-Lara, S., Prats, E., Sainz, J. and Cornudella, L. (1992). Cloning and characterization of a highly conserved satellite DNA from the mollusk *Mytilus edulis*. *Gene*, 117, 237-242.

Sadder, T. and Weber, G. (2002). Comparison between genetic and physical maps in *Zea mays* L. of molecular markers linked to resistance against *Diatraea spp*. *Theor. Appl. Genet.*, 104, 908-915.

Sahara, K., Yoshido, A., Marec, F. et al. (2007). Conserved synteny of genes between chromosome 15 of *Bombyx mori* and a chromosome of *Manduca sexta* shown by five-color BAC-FISH. *Genome*, 50, 1061-1065.

Saitoh, Y. and Ikeda, J-E. (1997). Chromosome microdissection and microcloning. *Chromosome Res.*, 5, 77-80.

Sajdak, S. L., Reed, K. M., and Phillips, R. B. (1998). Intraindividual and interspecies variation in the 5S rDNA of coregonid fish. *J. Mol. Evol.*, 46, 680-688.

Sakai, C., Konno, F., Nakano, O. et al. (2007). Chromosome elimination in the interspecific hybrid medaka between *Oryzias latipes* and *O. hubbsi*. *Chromosome Res.*, 15, 697-709.

Sallinen, R., Latvanlehto, A., Kvist, A. P. et al. (2001). Physical mapping of mouse collagen genes on Chromosome 10 by high-resolution FISH. *Mamm. Genome*, 12, 340-346.

Sánchez-Gea, J. F., Serrano, J. and Galián, J. (2000). Variability in rDNA loci in Iberian species of the genus *Zabrus* (Coleoptera: Carabidae) detected by fluorescence in situ hybridization. *Genome*, 43, 22-28.

Santini, S., Boore, J. L., and Meyer, A. (2003). Evolutionary conservation of regulatory elements in vertebrate Hox gene clusters. *Genome Res.*, 13, 1111-1122.

Santos, J., Serra, L., Solé, E. and Pascual, M. (2010). FISH mapping of microsatellite loci from *Drosophila subobscura* and its comparison to related species. *Chromosome Res.*, 18, 213-226.

Scalenghe, F., Turco, E., Edstron, J. E., Pirrota, V. and Melli, M. (1981). Microdissection and cloning of DNA from a specific region of *Drosophila melanogaster* polytene chromosomes. *Chromosoma,* 82, 205-216.

Schienman, J. E., Lozovskaya, E. R. and Strausbaugh, L. D. (1998). *Drosophila virilis* has atypical kinds and arrangements of histone repeats. *Chromosoma*, 107, 529-539.

Schlötterer, C. (2000). Evolutionary dynamics of microsatellite DNA. *Chromosoma*, 109, 365-371.

Schmid, M., Feichtinger, W., Nanda, I. et al. (1994). An extraordinary low diploid chromosome number in the reptile Gonatodes taniae (Squamata, Gekkonidae). *J. Hered.*, 85, 255-260.

Schmid, M., Ziegler, C. G., Steinlein, C., Nanda, I. and Schartl, M. (2006). Cytogenetics of the bleak (*Alburnus alburnus*), with special emphasis on the B chromosomes. *Chromosome Res.*, 14, 231-242.

Schueler, M. G., Higgins, A. W., Rudd, M. K., Gustashaw, K. and Willard, H. F. (2001). Genomic and genetic definition of a functional human centromere. *Science*, 294, 109-115.

Schwarzacher, T. (2003). DNA, chromosomes, and in situ hybridization. *Genome*, 46, 953-962.

Schwarzacher, T. and Heslop-Harrison P. (2000). Practical *in situ* hybridization (199 p.). Oxford: BIOS Scientific Publishers Limited.

Schwarzacher, T., Leitch, A. R., Bennett, M. D. and Heslop-Harrison, J. S. (1989). *In situ* localization of parental genomes in a wide hybrid. *Ann. Bot.*, 64, 315-324.

Senger, G., Lüdecke, H. J., Horsthemke, B. and Claussen, U. (1990). Microdissection of banded human chromosomes. *Hum. Genet.*, 84, 507-511.

Sessions, S. K. (2008). Evolutionary cytogenetics in salamanders. *Chromosome Res.*, 16, 183-201.

Shan, X., Ray, D. A., Bunge, J. A. and Peterson, D. G. (2009). A bacterial artificial chromosome library for the Australian saltwater crocodile *(Crocodylus porosus)* and its utilization in gene isolation and genome characterization. *BMC Genomics*, 10, S9.

Shapiro, J. A. and Sternberg, R. (2005). Why repetitive DNA is essential to genome function. *Biol. Rev.*, 80, 1-24.

Sharbel, T. F., Green, D. M. and Houben, A. (1998). B chromosome origin in the endemic New Zealand frog *Leiopelma hochstetteri* through sex chromosome devolution. *Genome*, 41, 14-22.

Shetty, S., Griffin, D. K. and Graves, J. A. M. (1999). Comparative painting reveals strong chromosome homology over 80 million years of bird evolution. *Chromosome Res.*, 7, 289-295.

Shi, L., Ye, Y. and Duan, X. (1980). Comparative cytogenetic studies on the red muntjac, Chinese muntjac and their F1 hybrids. *Cytogenet. Cell Genet.*, 26, 22-27.

Shibusawa, M., Nishibori, M., Nishida-Umehara, C. et al. (2004a). Karyotypic evolution in the Galliformes: an examination of the process of karyotypic evolution by comparison of the molecular cytogenetic findings with the molecular phylogeny. *Cytogenet. Genome Res.*, 106, 111-119.

Shibusawa, M., Nishida-Umehara, C., Tsudzuki, M. et al. (2004b). A comparative karyological study of the blue-breasted quail (*Coturnix chinensis*, Phasianidae) and California quail (*Callipepla californica*, Odontophoridae). *Cytogenet. Genome Res.*, 106, 82-90.

Sidow, A. (1996). Genome duplications in the evolution of early vertebrates. *Curr. Opin. Genet. Dev.*, 6, 715-722.

Siepel, A., Bejerano, G., Pedersen, J. S. et al. (2005). Evolutionarily conserved elements in vertebrate, insect, worm, and yeast genomes. *Genome Res.*, 15, 1034-1050.

Silva, M. J. J. and Yonenaga-Yassuda, Y. (1998). Heterogeneity and meiotic behavior of B and sex chromosomes, banding patterns and localization of (TTAGGG)n sequences by fluorescence *in situ* hybridization in the neotropical water rat *Nectomys* (Rodentia, Cricetidae). *Chromosome Res.*, 6, 455-462.

Silva, R. and Burch, J. B. E. (1989). Evidence that chicken CR1 elements represent a novel family of retroposons. *Mol. Cell. Biol.*, 9, 3563-3566.

Sinclair, A. H., Berta, P., Palmer, M. S. et al. (1990). A gene from the human sex-determining region encodes a protein with homology to a conserved DNA-binding motif. *Nature*, 346, 240-244.

Singh, L. (1972). Evolution of karyotypes in snakes. *Chromosoma*, 38, 185-236.

Singh, L., Purdom, I.F. and Jones, K.W. (1980). Sex chromosome associated satellite DNA: Evolution and conservation. *Chromosoma*, 79, 137-157.

Sittman, D. B., Chiu, I. M., Pan, C. J. et al. (1981). Isolation of two clusters of mouse histone genes. *Proc. Natl. Acad. Sci. U S A.*, 78, 4078-4082.

Slamovits, C. H. and Rossi, M. S. (2002). Satellite DNA: agent of chromosomal evolution in mammals. A review. *J. Neotrop. Mammal.*, 9, 297-308.

Slamovits, C. H., Cook, J. A., Lessa, E. P. and Rossi, M. S. (2001). Recurrent amplifications and deletions of satellite DNA accompanied chromosomal diversification in South American tucos-tucos (genus Ctenomys, Rodentia: Octodontidae): a phylogenetic approach. *Mol. Biol. Evol.*, 18, 1708-1719.

Smith, S. G. (1953). Chromosome numbers of Coleoptera. *Heredity*, 7, 31-48.

Smith, S. G. (1960). Cytogenetics of insects. *Annu. Rev. Entomol.*, 5, 69-84.

Sola, L., De Innocentiis, S., Gornung, E. et al. (2000). Cytogenetic analysis of *Epinephelus marginatus* (Pisces: Serranidae), with the chromosome localization of the 18S and 5S rRNA genes and of the (TTAGGG)n telomeric sequence. *Mar. Biol.,* 137, 47-51.

Solar, I. I., Hajen, W. E. and Donaldson, E. M. (1992). A bibliography of tetraploidy in fish (1964-1991). *Can. Tech. Rep. Fish. Aquat. Sci.*, 1901, 22.

Solovei, I., Ogawa, A., Naito, M., Mizuno, S., and Macgregor, H. (1998). Specific chromomeres on the chicken W lampbrush chromosome contain specific repetitive DNA sequence families. *Chromosome Res.*, 6, 323-327.

Speicher, M. R. and Carter, N. P. (2005). The new cytogenetics: Blurring the boundaries with molecular biology. *Nat. Rev. Genet.*, 6, 782-792.

Srikulnath, K., Nishida, C., Matsubara, K. et al. (2009). Karyotypic evolution in squamate reptiles: comparative gene mapping revealed highly conserved linkage homology between the butterfly lizard (*Leiolepis reevesii rubritaeniata*, Agamidae, Lacertilia) and the Japanese four-striped rat snake (*Elaphe quadrivirgata*, Colubridae, Serpentes). *Chromosome Res.,* 17, 975-986.

Srivastava, M., Begovic, E., Chapman, J. et al. (2008). The *Trichoplax* genome and the nature of placozoans. *Nature*, 454, 955-960.

Stace, C. A. and Bailey, J. P. (1999). The value of genomic insitu hybridization (GISH) in plant taxonomic and evolutionary studies. In P. M. Hollingsworth, R. M. Bateman and R. J. Gornall (eds.), Molecular systematics and plant evolution (pp. 199-210). London: Taylor and Francis limited.

Stack, S. M. (1984). Heterochromatin, the synaptonemal complex and crossing over. *J. Cell Sci.*, 71, 159-176.

Stanyon, R., Rocchi, M., Capozzi, O. et al. (2008). Primate chromosome evolution: Ancestral karyotypes, marker order and neocentromeres. *Chromosome Res.*, 16, 17-39.

Stanyon, R., Yang, F., Cavagna, P. et al. (1999). Reciprocal chromosome painting shows that genomic rearrangement between rat and mouse proceeds ten times faster than between humans and cats. *Cytogenet. Cell Genet.*, 84, 150-155.

Stein, J., Phillips, R. B. and Devlin, R. H. (2001). Identification of the Y chromosome in chinook salmon (*Oncorhynchus tshawytscha*). *Cytogenet. Cell Genet.*, 92, 108-110.

Stiglec, R., Ezaz, T. and Graves, J. A. M. (2007a). Reassignment of chicken W chromosome sequences to the Z chromosome by fluorescence *in situ* hybridization (FISH). *Cytogenet. Genome Res.*, 116, 132-134.

Stiglec, R., Ezaz, T. and Graves, J. A. M. (2007b). A new look at the evolution of avian sex chromosomes. *Cytogenet. Genome Res.*, 117, 103-109.

Stitou, S., de la Guardia, D., Jiménez, R. and Burgos, M. (2000). Inactive ribosomal cistrons are spread throughout the B chromosomes of *Rattus rattus* (Rodentia, Muridae). Implications for their origin and evolution. *Chromosome Res.,* 8, 305-311.

Stock, A. D. and Bunch, T. D. (1982). The evolutionary implications of chromosome banding pattern homologies in the bird order Galliformes. *Cytogenet. Cell Genet.*, 34, 136-148.

Strasburger, E. (1875). *Über zellbildung und zelltheilung.* (1st edition). Jena: Hermann Dabis.

Stumph, W. E., Hodgson, C. P., Tsai, M. J. and O'Malley, B. W. (1984). Genomic structure and possible retroviral origin of the chicken CR1 repetitive DNA sequence family. *Proc. Natl. Acad. Sci. USA*, 81, 6667-6671.

Stumph, W. E., Kristo, P., Tsai, M. J. and O'Malley, B. W. (1981). A chicken middle-repetitive DNA sequence which shares homology with mammalian ubiquitous repeats. *Nucleic Acids Res.*, 9, 5383–5397.

Subirana, J. A. and Messeguer, X. (2010). The most frequent short sequences in non-coding DNA. *Nucleic Acids Res.*, 38, 1172-1181.

Sugawara, N. and Szostak, J. W. (1986). Telomeres of *Schizosaccharomyces pombe*. *Yeast*, 2, S373.

Sumner, A. T. (1972). A simple technique for demonstrating centromeric heterochromatin. *Exp. Cell. Res.*, 75, 304-306.

Sumner, A. T. (1990). Chromosome banding. Unwin Hyman, London: Unwin Hyman.

Sumner, A.T. (2003). Chromosomes: organization and function (1st edition). London: Blackwell Publishing company.

Sun, F., Oliver-Bonet, M., Liehr, T. and et al. (2004). Human male recombination maps for individual chromosomes. *Am. J. Hum. Genet.*, 74, 521-531.

Suto, Y., Ishikawa, Y., Hyodo, H. et al. (2003). Gene arrangement at the Rhesus blood group locus of chimpanzees detected by fiber-FISH. *Cytogenet. Genome Res.*, 101, 161-165.

Sutton, W. S. (1902). On the morphology of the chromosome group in *Brachystola magna*. *Biol. Bull*, 4, 24-39.

Svartman, M. and Vianna-Morgante, A. M. (1999). Comparative genome analysis in American marsupials: chromosome banding and insitu hybridization. *Chromosome Res.*, 7, 267-275.

Svartman, M., Stone, G. and Stanyon, R. (2004a). Molecular cytogenetics discards polyploidy in mammals. *Genomics*, 85, 425-430.

Svartman, M., Stone, G., Page, J. E. and Stanyon, R. (2004b). A chromosome painting test of the basal eutherian karyotype. *Chromosome Res.*, 12, 45-53.

Swinger, R. R. and Tucker, J. D. (1996). Fluorescence *in situ* hybridization: a brief review. *Environ. Mol. Mutagen*, 27, 245-254.

Sybenga, J. (1999). What makes homologous chromosomes find each other in meiosis? A review and an hypothesis. *Chromosoma*, 108, 209-219.

Syvanen, M. and Ducore, J. (2010). Whole genome comparisons reveals a possible chimeric origin for a major metazoan assemblage. *J. Biol. Systems*, 18, 261-275.

Tanaka, K., Suzuki, T., Nojiri, T. et al. (2000). Characterization and chromosomal distribution of a novel satellite DNA sequence of Japanese quail (*Coturnix coturnix japonica*). *J. Hered.*, 91, 412-415.

Tatsuke, T., Sakashita, K., Masaki, Y., et al. (2010). The telomere-specific non-LTR retrotransposons SART1 and TRAS1 are suppressed by piwi subfamily proteins in the silkworm, *Bombyx mori. Cell. Mol. Biol. Lett.*, 11, 118-133.

Tay, W. T., Behere, G., Batterham, P. and Heckel, D. G. (2010). Generation of microsatellite repeat families by RTE retrotransposons in lepidopteran genomes. *BMC Evol. Biol.*, 10, 144.

Teixeira, W. G., Ferreira, I. A., Cabral-de-Mello, D. C. et al. (2009). Organization of repeated DNA elements in the genome of the cichlid fish *Cichla kelberi* and its contributions to the knowledge of fish genomes. *Cytogenet. Genome Res.*, 125, 224-234.

Telenius, H., Carter, N. P., Bebb, C. E. et al. (1992). Degenerate oligonucleotide-primed PCR: general amplification of target DNA by a single degenerate primer. *Genomics*, 13, 718-725.

Terkelsen, C., Koch, J., Kolvraa, S. et al. (1993). Repeated primed *in situ* labeling: formation and labeling of specific DNA sequences in chromosomes and nuclei. *Cytogenet. Cell Genet.*, 63, 235-237.

Teruel, M., Cabrero, J., Montiel, E. E. et al. (2009a). Microdissection and chromosome painting of X and B chromosomes in *Locusta migratoria*. *Chromosome Res.*, 17, 11-18.

Teruel, M., Cabrero, J., Perfectti, F. et al. (2009b). Microdissection and Chromosome Painting of X and B Chromosomes in the Grasshopper *Eyprepocnemis plorans*. *Cytogenet. Genome Res.*, 125, 286-291.

Teruel, M., Cabrero, J., Perfectti, F. and Camacho, J. P. M. (2010). B chromosome ancestry revealed by histone genes in the migratory locust. *Chromosoma*, 119, 217-225.

Tharapel, A. T., Kadandale, J. S., Martens, P. R., Wachtel, S. S. and Wilroy Jr., R. S. (2002). Prader Willi/Angelman and DiGeorge/Velocardiofacial Syndrome Deletions: Diagnosis by Primed *In situ* Labeling (PRINS). *Am. J. Med. Genet.*, 107, 119-122.

The *Caenorhabditis elegans* Sequencing Consortium (1998). Genome sequence of the nematode *C. elegans*: a platform for investigating biology. *Science*, 282, 2012-2018.

Theuns, J., Cruts, M., Del-Favero, J. et al. (1999). Determination of the genomic organization of human presenilin 1 by fiber-FISH analysis and restriction mapping of cloned DNA. *Mamm. Genome*, 10, 410-414.

Thomas, R., Duke, S. E., Bloom, S. K. et al. (2007). A cytogenetically characterized, genome-anchored 10-Mb BAC set and CGH array for the domestic dog. *J. Hered.*, 98, 474-484.

Thomas, R., Valli, V. E., Ellis, P. et al. (2009). Microarray-based cytogenetic profiling reveals recurrent and subtype-associated genomic copy number aberrations in feline sarcomas. *Chromosome Res.*, 17, 987-1000.

Thompson, J. D., Higgins, D. G. and Gibson, T.J. (1994). CLUSTAL W: improving the sensitivity of progressive multiple sequence alignment through sequence weighting, position-specific gap penalties and weight matrix choice. *Nucleic Acids Res.*, 22, 4673-4680.

Thorsen, J., Zhu, B., Frengen, E. et al. (2005). A highly redundant BAC library of Atlantic salmon (*Salmo salar*): an important tool for salmon projects. *BMC Genomics*, 6, 50.

Tjio, J. H. and Levan, A. (1956). The chromosome number of man. *Hereditas*, 42, 1-6.

Tobler, H., Etter, A. and Muller, A. (1992). Chromatin diminution in nematode development. *Trends Genetic.*, 8, 427-432.

Todd, N. B. (1970). Karyotypic fissioning and canid phylogeny. *J. Theor. Biol.*, 26, 445-480.

Toder, R., Xia, Y. and Bausch, E. (1998). Interspecies comparative genome hybridization and interspecies representational difference analysis reveal gross DNA differences between humans and great apes. *Chromosome Res.*, 6, 487-494.

Tör, M., Manning, K., King, G. J. et al. (2002). Genetic analysis and FISH mapping of the colorless non-ripening locus of tomato. *Theor. Appl. Genet.*, 104, 165-170.

Traut, W., Eickhoff, U. and Schorch, J (2001). Identification and analysis of sex chromosomes by comparative genomic hybridization (CGH). *Methods Cell Sci.*, 23, 155-161.

Traut, W., Weichenhan, D., Eickhoff, U. and Winking, E. (1999a). Origin of the chromosome 1 HSR of the house mouse detected by CGH. *Chromosome Res.*, 7, 649-653.

Traut, W., Sahara, K., Otto, T. D. and Marec, F. (1999b). Molecular differentiation of sex chromosomes probed by comparative genomic hybridization. *Chromosoma*, 108, 173-180.

Trifonov, V. A., Perelman, P. L., Kawada, S. I. et al. (2002). Complex structure of B chromosomes in two mammalian species: *Apodemus peninsulae* (Rodentia) and *Nyctereutes procyonoides* (Carnivora). *Chromosome Res.*, 10, 109-116.

Trifonov, V. A., Stanyon, R., Nesterenko, A. I. et al. (2008). Multidirectional cross-species painting illuminates the history of karyotypic evolution in Perissodactyla. *Chromosome Res.*, 16, 89-107.

Tripputi, P., Emanuel, B. S., Croce, C. M., Green, L.G., Stein, G. S. and Stein, J. L. (1986). Human histone genes map to multiple chromosomes. *Proc. Natl. Acad. Sci. USA*, 10, 3185-3188.

Trivedi, S. (2004). Microsatellites (SSRs): puzzles within puzzle. *Indian J. Biotechnol.*, 3, 331-347.

Tsunemoto, K. and Matsuo, Y. (2001). Molecular evolutionary analysis of a histone gene repeating unit from *Drosophila simulans*. *Genes Genet. Syst.*, 76, 355-361.

Turner, P. C., Bagenal, E. B. D., Vlad, M. T. and Woodland H. R. (1988) The organization and expression of histone genes from *Xenopus borealis*. *Nucleic Acids Res.*, 16, 3171-3485.

Ueshima, N. (1979). Hemiptera II: Heteroptera. In B. John (ed.) Animal cytogenetics (117 p.). Berlin: Gebrüder Borntraeger.

Ugarković, D. and Plohl, M. (2002). Variation in satellite DNA profiles-causes and effects. *EMBO J.*, 21, 5955-5959.

Uller, T., Pen, I., Wapstra, E., Beukeboom, L. W. and Komdeur, J. (2007). The evolution of sex ratios and sex-determining systems. *Trends Ecol. Evol.*, 22, 292-297.

Vafa, O., Shelby, R. D. and Sullivan, K. F. (1999). CENP-A associated complex satellite DNA in the kinetochore of the Indian muntjac. *Chromosoma*, 108, 367-374.

Valente, G. T., Schneider, C. H., Gross, M. C., Feldberg, E. and Martins, C. (2009). Comparative cytogenetics of cichlid fishes through genomic in-situ hybridization (GISH) with emphasis on *Oreochromis niloticus*. *Chromosome Res.*, 17, 791-799.

Valente, G. T., Mazzuchelli, J., Ferreira, I. A., Poletto, A. B., Fantinatti, B. E. A., Martins, C. (2010). Cytogenetic mapping of the retroelements *Rex1*, *Rex3* and *Rex6* among cichlid fish: new insights on the chromosomal distribution of transposable elements. *Cytogenet. Genome Res.*, (in press).

Van de Peer, Y., and Meyer, A. (2005). Large-scale gene and ancient genome duplications. In T. R. Gregory *(Ed.), The Evolution of the Genome* (pp. 330-363). San Diego: Elsevier.

Veltsos, P., Keller, I. and Nichols, R. A. (2009). Geographical localized bursts of ribosomal DNA mobility in the grasshopper *Podisma pedestris*. *Heredity*, 103, 54-61.

Ventura, K., O'Brien, P. C. M., Yonenaga-Yassuda, Y. and Ferguson-Smith, M. A. (2009). Chromosome homologies of the highly rearranged karyotypes of four Akodon species (Rodentia, Cricetidae) resolved by reciprocal chromosome painting: the evolution of the lowest diploid number in rodents. *Chromosome Res.*, 17, 1063-1078.

Vergnaud, G. and Denoeud, F. (2000). Minisatellites: Mutability and Genome Architecture. *Genome Res.*, 10, 899-907.

Verhulst, E. C., van de Zande, L. and Beukeboom, L. W. (2010). Insect sex determination: it all evolves around transformer. *Curr. Opin. Genet. Devel.*, 20, 376-383.

Veyrunes, F., Waters, P. D., Miethke, P. et al. (2008). Bird-like sex chromosomes of platypus imply recent origin of mammal sex chromosomes. *Genome Res.*, 18, 965-973.

Viera, A., Calvente, A., Page, J. et al. (2004). X and B chromosomes display similar meiotic characteristics in male grasshoppers. *Cytogenet. Genome Res.*, 106, 302-308.

Villasante, A., de Pablos, B., Méndez-Lago, M. and Abad, J. P. (2008). Telomere maintenance in *Drosophila*: rapid transposon evolution to chromosome ends. *Cell Cycle*, 7, 2134-2138.

Vinãs, A., Abuin, M., Pardo, B. G., Martí P. And Sánchez, L. (2004). Characterization of a new HpaI centromeric satellite DNA in *Salmo salar*. *Genetica*, 121, 81-87.

Vítková, M., Fuková, I., Kubíčková, S. and Marec, F. (2007). Molecular divergence of the W chromosomes in pyralid moths (Lepidoptera). *Chromosome Res.*, 15, 917-930.

Vitturi, P. R., Libertini, A., Armetta, F., Sparacino, L. and Colomba, M. S. (2002). Chromosome analysis and FISH mapping of ribosomal DNA (rDNA), telomeric (TTAGGG)n and (GATA)n repeats in the leech Haemopis sanguisuga (L.) (Annelida: Hirudinea). *Genetica*, 115, 189-194.

Vitturi, R., Sineo, L., Volpe, N., Lannino, A. and Colomba, M. (2004). Repetitive DNAs in the slug *Milax nigricans*: association of ribosomal (18S–28S and 5S rDNA) and (TTAGGG)n telomeric sequences) in the slug *M. nigricans* (Mollusca: Gastropoda: Pulmonata). *Micron*, 35, 255-260.

Volff, J-N., Bouneau, L., Ozouf-Costaz, C. and Fischer, C. (2003). Diversity of retrotransposable elements in compact pufferfish genomes. *Trends Genet.*, 19, 674-678.

Völker, M., Backström, N., Skinner, B. M. et al. (2010). Copy number variation, chromosome rearrangement, and their association with recombination during avian evolution. *Genome Res.*, 20, 503-511.

Volleth, M., Heller, K-G., Pfeiffer, R. A. and Hameister, H. (2002). A comparative Zoo-FISH analysis in bats elucidates the phylogenetic relationships between Megachiroptera and five microchiropteran families. *Chromosome Res.*, 10, 477-497.

Volpi, E. V. and Bridger, J. M. (2008). FISH glossary: an overview of the fluorescence *in situ* hybridization technique. *Biotechniques*, 45, 385-409.

Wachtler, F., Stahl, A. (1993). The nucleolus: a structural and functional interpretation. *Micron*, 24, 473-505.

Wahls, W. P. and Moore, P. D. (1998). Recombination hotspot activity of hypervariable minisatellite DNA requires minisatellite DNA binding proteins. *Somat. Cell Mol. Genet.*, 24, 41-51.

Waldeyer, Q. (1890). Karyokinesis and its relation to the process of fertilization. *Q. J. Micr. Sci.*, 30, 159-281.

Wallis, M. C., Waters, P. D. and Graves J. A. M. (2008). Sex determination in mammals-Before and after the evolution of SRY. *Cell. Mol. Life Sci.*, 65, 3182-3195.

Wan, T. S. K., Martens, U. M., Poon, S. S. S. et al. (1999). Absence or Low Number of Telomere Repeats at Junctions of Dicentric Chromosomes. *Genes Chromosomes Cancer*, 24, 83-86.

Wang, S., Zhang, L., Meyer, E. and Bao, Z. (2010). Genomes-wide analysis of transposable elements and tandem repeats in the compact placozoan genome. *Biol. Dir.*, 5, 1-9.

Wang, Y. and Guo, X. (2004). Chromosomal rearrangement in Pectinidae revealed by rRNA loci and implications for bivalve evolution. *Biol. Bull*, 207, 247-256.

Wang, Y., Xu, Z. and Guo, X. (2001). A centromeric satellite sequence in the Pacific oyster (*Crassostrea gigas* Thunberg) identified by fluorescence *in situ* hybridization. *Mar. Biotechnol.* 3, 486-492.

Wanner, G. and Formanek, H. (2000). A new chromosome model. *J. Struct. Biol.*, 132, 147-61.

Warren, W. C., Hillier, L. W., Graves, J. A. M. et al. (2008). Genome analysis of the platypus reveals unique signatures of evolution. *Nature*, 453, 175-183.

Watabe, H., Bachmann, L., Haring, E. and Sperlich, D. (1997). Taxonomic and molecular studies on *Drosophila sinobscura* and *D. hubeiensis*, two sibling species of the *D. obscura* group. *J. Zool. Syst. Evol. Res.*, 35, 81-94.

Waterston, R. H., Lindblad-Toh, K., Birney, E., et al. (2002). Initial sequencing and comparative analysis of the mouse genome. *Nature*, 420, 520-562.

Watson, J. D. and Crick, F. H. (1953). Molecular structure of nucleic acids; a *structure for deoxyribose* nucleic acid. *Nature*, 171, 737-738.

Weier, H. U. G. (2001). DNA fiber mapping techniques for the assembly of high-resolution physical maps. *J. Hist. Cytoch.*, 49, 939-948.

Weller, P., Jeffreys, A. J., Wilson, V. and Blanchetot, A. (1984). Organization of the human myoglobin gene. *EMBO J.*, 3, 439-446.

Werner, P., Mellersh, C. S., Raducha, M. G. et al. (1999). Anchoring of canine linkage groups with chromosome-specific markers. *Mamm. Genome*, 10, 814-823.

Werren, J. H. and Stouthamer, R. (2003). PSR (paternal sex ratio) chromosomes: the ultimate selfish genetic elements. *Genetica*, 117, 85-101.

Wesley, C.S., Ben, M., Kreitman, M., Hagag, N. and Eanes, W. F. (1990). Cloning regions of the *Drosophila* genome by microdissection of polytene chromosome DNA and PCR with nonspecific primer. *Nucleic Acids Res.*, 18, 599-603.

White, M. J. D. (1973). Animal cytology and evolution (3rd ed.). London: Cambridge University.

Wichman, H. A., Payne, C. T., Ryder, O. A., et al. (1991). Genomic distribution of heterochromatic sequences in equids: implications to rapid chromosomal evolution. *J. Hered*, 82, 369-377.

Wichman, H. A., Van den Bussche, R. A., Hamilton, M. J. & Baker, R. J. (1992). Transposable elements and the evolution of genome organization in mammals. *Genetica*, 86, 287-293.

Wicker, T., Robertson, J. S., Schulze, S. R., et al. (2005). The repetitive landscape of the chicken genome. *Genome Res.*, 15, 126-136.

Wicker, T., Sabot, F., Hua-Van, A., et al. (2007). A unified classification system for eukaryotic transposable elements. *Nature Rev. Genet.*, 8, 973-982.

Wienberg, J. (2004). The evolution of eutherian chromosomes. *Curr. Opin. Genet. Dev.*, 14, 657-666.

Wienberg, J., Jauch, A., Stanyon, R. and Cremer, T. (1990). Molecular cytotaxonomy of primates by chromosomal *in situ* suppression hybridization. *Genomics*, 8, 347-350.

Wilkens, L., Tchinda, J., Komminoth, P. and Werner, M. (1997). Single- and double-color oligonucleotide primed *in situ* labeling (PRINS): applications in pathology. *Histochem. Cell Biol.*, 108, 439-446.

Wilkinson, D.G. (1999). The theory and practice of *in situ* hybridization. In D.G. Wilkinson (Ed.), *In situ* hybridization: A practical approach. (2nd ed., pp. 1-22). New York: Oxford University Press.

Willard, H. F. and Waye, J. S. (1987). Chromosome-specific subsets of human alpha satellite DNA: Analysis of sequence divergence within and between chromosomal subsets and evidence for an ancestral pentameric repeat. *J. Mol. Evol.*, 25, 207-214.

Willhoeft, U., Mueller-Navia, J. and Franz, G. (1998). Analysis of the sex chromosomes of the Mediterranean fruit fly by microdissected DNA probes. *Genome*, 41, 74-78.

Wilson, E. B. (1907a). The supernumerary chromosomes of Hemiptera. *Science*, 26, 870-871.

Wilson, E. B. (1907b). Note on the chromosome groups of *Metapodius* and *Banasa*. *Biol. Bull.*, XII, 303-313.

Winkler, H. (1920). *Verbreitung und ursache der parthenogenesis im pflanzen- und tierreiche*. Jena: Fischer Verlag.

Wnuk, M., Bugno, M. and Slota, E. (2008). Application of primed *in situ* DNA synthesis (PRINS) with telomere human commercial kit in molecular cytogenetics of *Equus caballus* and Sus scrofa scrofa. *Folia Histochem. Cytobiol.*, 46, 85-88.

Wolf, K. W., Mertl, H. G. and Traut, W. (1991). Structure, mitotic and meiotic behavior, and stability of centromere-like elements devoid of chromosome arms in the fly *Megaselia scalaris* (Phoridae). *Chromosoma*, 101, 99-108.

Wolffe, A. P. (1999). Chromatin: Structure and Function, (3rd ed. 447 p.). San Diego: Academic Press.

Woodhouse, M., Burkart-Waco, D. and Comai, L. (2009). Polyploidy. *Nat. Educ.*, 2, 1.

Woods, I. G., Wilson, C., Friedlander, B., et al. (2005). The zebrafish gene map defines ancestral vertebrate chromosomes. *Genome Res.*, 15, 1307-1314.

Wrigley, J. M., Graves, J. A. M. (1988). Karyotypic conservation in the mammalian order monotremata (subclass Prototheria). *Chromosoma*, 96, 231-247.

Wu, C., Proestou, D., Carter, D., et al. (2009). Construction and sequence sampling of deep-coverage, large-insert BAC libraries for three model lepidopteran species. *BMC Genomics*, 10, 283.

Wurster, D. H. and Benirschke, K. (1967). Chromosome studies in some deer, the springbok and the pronghorn, with notes on placentation in deer. *Cytologia*, 32, 273-285.

Wurster, D. H. and Benirschke, K. (1968). Comparative cytogenetic studies of the order carnivora. *Chromosoma*, 24, 336-382.

Wurster, D. H. and Benirschke, K. (1970). Indian Muntjac, *Muntiacus muntjak*: a deer with a low diploid number. *Science*, 168, 1364-1366.

Wurster-Hill, D. H., Ward, O. G., Davis, B. H., et al. (1988). Fragile sites, telomeric DNA sequences, B chromosomes, and DNA content in raccoon dogs, *Nyctereutes procyonides*, with comparative notes on foxes, coyote, wolf and raccoon. *Cytogenet. Cell Genet.*, 49, 278-281.

Xiong, Y. and Eickbush, H. (1990). Origin and evolution of retroelements based upon their reverse transcriptase sequences. *EMBO J.*, 9, 3353-3362.

Yamada, K., Kamimura, E., Kondo, M., et al. (2006). New families of site-specific repetitive DNA sequences that comprise constitutive heterochromatin of the Syrian hamster (*Mesocricetus auratus*, Cricetinae, Rodentia). *Chromosoma*, 115, 36-49.

Yamada, K., Nishida-Umehara, C., Matsuda, Y. (2002a). Characterization and chromosomal distribution of novel satellite DNA sequences of the lesser rhea (*Pterocnemia pennata*) and the greater rhea (*Rhea americana*). *Chromosome Res.*, 10, 513-523.

Yamada, K., Shibusawa, M., Tsudzuki, M. and Matsuda, Y. (2002b). Molecular cloning and characterization of novel centromeric repetitive DNA sequences in the blue-breasted quail (*Coturnix chinensis*, Galliformes). *Cytogenet Genome Res.*, 98, 255-261.

Yang, F. and Graphodatsky, A. S. (2009). Animal Probes and Zoo-Fish. In T. Liehr (ed.), Fluorescence in Situ Hybridization (Fish): Application Guide (pp. 323-346). Berlin: Springer.

Yang, F., Alkalaeva, E. Z., Perelman, P. L., et al. (2003). Reciprocal chromosome painting among human, aardvark, and elephant (superorder Afrotheria) reveals the likely eutherian ancestral karyotype. *Proc. Natl. Acad. Sci. USA*, 100, 1062-1066.

Yang, F., Carter, N. P., Shi, L. and Ferguson-Smith, M. A. (1995). A comparative study of karyotypes of muntjacs by chromosome painting. *Chromosoma*, 103, 642-652.

Yang, F., Graphodatsky, A. S., Li, T., et al. (2006). Comparative genome maps of the pangolin, hedgehog, sloth, anteater, and human revealed by cross-species chromosome painting: Further insight into the ancestral karyotype and genome evolution of eutherian mammals. *Chromosome Res.*, 14, 283-296.

Yang, F., Müller, S., Just, R., Ferguson-Smith, M. A. and Wienberg, J. (1997a). Comparative chromosome painting in mammals: human and the Indian muntjac (*Muntiacus muntjak vaginalis*). *Genomics*, 39, 396-401.

Yang, F., O'Brien, P. C. M., Milne, B. S., et al. (1999). A complete comparative chromosome map for the dog, red fox and human and its integration with canine genetic maps. *Genomics*, 62, 189-202.

Yang, F., O'Brien, P. C. M., Wienberg, J. and Ferguson-Smith M. A. (1997b). A reappraisal of the tandem fusion theory of karyotype evolution in the Indian muntjac using chromosome painting. *Chromosome Res.*, 5, 109-117.

Yang, F., Trifonov, V. A., Ng, B.L., Kosyakova, N. and Carter, N. P. (2009). Generation of paint probes by flow-sorted and microdissected chromosomes. In T. Liehr (eds.), Fluorescence *in situ* hybridization (FISH): application guide. (pp. 35-52). Berlin: Springer.

Yasukochi, Y., Tanaka-Okuyama, M., Shibata, F., et al. (2009). Extensive conserved synteny of genes between the karyotypes of *Manduca sexta* and *Bombyx mori* revealed by BAC-FISH mapping. *PLoS One*, 4, e7465.

Ye, J., Biltueva, L., Huang, L., et al. (2006). Cross-species chromosome painting unveils cytogenetic signatures for the Eulipotyphla and evidence for the polyphyly of Insectivora. *Chromosome Res.*, 14, 151-159.

Yoshido, A., Bando, H., Yasukochi, Y. and Sahara, K. (2005). The *Bombyx mori* karyotype and the assignment of linkage groups. *Genetics*, 170, 675-685.

Yoshimoto, S., Okada, E., Umemoto, H., et al. (2008). A W-linked DM-domain gene, DM-W, participates in primary ovary development in *Xenopus laevis. Proc. Natl. Acad. Sci. USA*, 105, 2469-2474.

Yudkin, D. V., Trifonov, V. A., Kukekova, A. V., et al. (2007). Mapping of KIT adjacent sequences on canid autosomes and B chromosomes. *Cytogenet. Genome Res.*, 116, 100-103.

Yunis, J. J. and Yasmineh, W. G. (1971). Heterochromatin, satellite DNA, and cell function. *Science,* 174, 1200-1209.

Zdobnov, E. M., von Mering, C., Letunic, I., et al. (2002). Comparative genome and proteome analysis of *Anopheles gambiae* and *Drosophila melanogaster*. *Science,* 298, 149-159.

Zhang, L., Bao, Z., Wang, S., Huang, X. and Hu, J. (2007). Chromosome rearrangements in Pectinidae (Bivalvia: Pteriomorphia) implied based on chromosomal localization of histone H3 gene in four scallops. *Genetica,* 130, 193-198.

Zickler, D. and Kleckner, N. (1999). Meiotic chromosomes: integrating structure and function. *Annu. Rev. Genet.*, 33, 603-754.

Ziegler, C. G., Lamatsch, D. K., Steinlein, C., Engel, W., Schartl, M. and Schmid, M. (2003). The giant B chromosome of the cyprinid fish *Alburnus alburnus* harbours a retrotransposon-derived repetitive DNA sequence. *Chromosome Res.,* 11, 23-35.

Zwick, M. S., Hanson, R. E., McKnight, T. D., Nurul-Islam-Faridi, M. and Stelly, D. M. (1997). A rapid procedure for the isolation of C_0t-1 DNA from plants. *Genome,* 40, 138-142.

INDEX

A

access, 39, 75
accessibility, 76, 77
acetylation, 3, 4, 6, 54
acid, 16, 148
acrocentric chromosome, 10, 13, 14, 61
adaptation, 118, 139
adaptations, 118
adenine, 31
ADP, 54
age, 10, 14, 114, 130
allopatric speciation, 125
allopolyploid, 44
alters, 25
amino, 6
amino acid, 6
amphibia, 8, 67, 68, 95
amphibians, 15, 16, 18, 19, 22, 45, 52, 58, 59, 67, 89, 91, 92, 95, 97, 128, 130, 134
anatomy, 128
ancestors, 22
anchoring, 6, 15
androgens, 73
aneuploid, 23
aneuploidy, 27, 44, 99
annealing, 31, 42, 71, 73
annotation, 27
antibody, 42
Arabidopsis thaliana, 56, 126
arthropods, 52, 62, 66
ATP, 6
avian, 13, 18, 24, 62, 68, 75, 76, 86, 87, 92, 93, 94, 120, 122, 123, 125, 126, 134, 140, 143, 147
avoidance, 68

B

BAC, 32, 33, 35, 37, 39, 40, 46, 71, 72, 73, 74, 75, 79, 81, 82, 88, 96, 111, 115, 124, 126, 128, 133, 139, 141, 145, 149, 150
bacteria, 26, 124
bacterial artificial chromosome, 2, 39, 71, 73, 115, 121, 125, 129, 142
barriers, 11
base, 4, 18, 19, 23, 24, 27, 29
base pair, 4, 19, 23, 24
basic research, 43, 75
beetles, 9, 14, 53, 55, 57, 127
benefits, 99
bias, 21, 56
biochemistry, 1
biodiversity, 111
bioinformatics, 2, 29, 30, 71, 84, 88, 103, 106, 120
biological consequences, 26
biological processes, 106
biotin, 41
birds, 13, 15, 18, 37, 39, 40, 41, 58, 61, 62, 67, 68, 75, 79, 80, 86, 87, 88, 92, 93, 94, 95, 97, 98, 113, 116, 118, 131, 135
blood, 26, 65, 77, 144
blood group, 77, 144
brain, 123
Brazil, 8, 9, 17, 45
breast cancer, 105
breeding, 139
budding, 3
buffalo, 77

C

cancer, 7, 44, 73, 105
cancer cells, 44, 105

candidates, 128
carbon, 41
carnivores, 81
cartilaginous, 95
catfish, 72, 139
cattle, 21, 73, 77, 82, 121, 126
Caucasus, 101
cDNA, 39, 40
cell culture, 19, 29, 35, 39
cell cycle, 6, 16, 19
cell division, 1, 4, 6
cell size, 132
central nervous system, 121
centric fusion, 14, 60, 82
centromere, 6, 14, 24, 34, 35, 36, 60, 66, 75, 76, 77, 99, 115, 119, 123, 124, 130, 141, 149
chemical, 4, 41, 71, 76, 119
chicken, 57, 62, 67, 68, 72, 73, 74, 75, 85, 86, 87, 92, 93, 94, 109, 116, 117, 118, 120, 122, 123, 124, 125, 127, 130, 131, 134, 138, 140, 142, 143, 144, 148
children, 105
chimpanzee, 82, 129
chromatid, 6, 7
chromosome 10, 115
chromosome bands, 77, 80
chromosome map, 29, 40, 72, 115, 119, 121, 150
classes, 28, 50, 53, 63, 64, 67, 116
classical methods, 20
classification, 55, 63, 64, 126, 131, 148
cleavage, 36
clinical trials, 105
clone, 73, 74, 79, 115
cloning, 35, 36, 39, 44, 72, 77, 100, 131, 141, 150
clusters, 25, 45, 53, 54, 55, 56, 65, 70, 109, 110, 114, 128, 134, 141, 142
coding, 22, 24, 25, 26, 62, 69, 73, 111, 129, 144
codon, 21
collagen, 76, 141
commercial, 72, 149
compaction, 6
comparative analysis, 16, 18, 19, 27, 62, 73, 79, 101, 106, 121, 148
complement, 13, 15, 44, 48, 58, 59, 79, 80, 98, 100, 113
complementary DNA, 31
complexity, 21, 24, 27, 35, 39, 49, 130
composition, 18, 19, 21, 59, 102, 129
comprehension, 3
computer, 47
condensation, 3, 5, 6, 57, 75, 102, 112
condor, 68
configuration, 3, 92, 93

connectivity, 11
consensus, 4, 49, 67, 115
conservation, 11, 14, 24, 53, 55, 56, 57, 60, 62, 73, 81, 86, 93, 94, 95, 96, 97, 101, 110, 134, 141, 142, 149
construction, 32, 35, 37, 39, 72, 73, 88
convergence, 86
correlation, 24, 59, 115, 139
cost, 21, 35, 69
cotton, 113
crocodile, 95, 126, 142
crossing over, 27, 143
cure, 105
cycles, 42, 74
cycling, 74
cytochrome, 128
cytogenetics, iv, 1, 2, 3, 15, 19, 20, 29, 30, 31, 32, 48, 72, 73, 80, 84, 88, 98, 105, 106, 107, 112, 114, 120, 121, 122, 130, 133, 136, 141, 143, 144, 146
cytology, 148
cytometry, 38, 120, 121, 122
cytosine, 31

D

data set, 30
database, 21, 73, 136
decay, 42, 109
defects, 7
degenerate, 38, 144
denaturation, 31, 36, 42
deoxyribose, 148
detachment, 43
detectable, 70, 71
detection, 15, 19, 31, 39, 41, 44, 71, 74, 75, 76, 77, 112, 120, 134, 136, 138
devolution, 142
diabetes, 105
diakinesis, 14
digestion, 115
digital cameras, 47
diploid, 2, 8, 10, 11, 12, 13, 14, 22, 23, 29, 62, 83, 87, 90, 98, 141, 146, 149
diplotene, 75
discordance, 69
discrimination, 44
diseases, 49, 72, 105
distribution, 15, 34, 53, 56, 57, 58, 59, 61, 62, 65, 67, 68, 69, 71, 76, 110, 111, 112, 113, 115, 119, 124, 128, 129, 130, 131, 132, 136, 138, 144, 146, 148, 150
divergence, 11, 27, 30, 67, 80, 82, 84, 86, 87, 93, 97, 101, 103, 106, 147, 149

diversification, 1, 10, 16, 59, 60, 62, 65, 86, 111, 142
diversity, 12, 14, 21, 24, 25, 27, 52, 54, 62, 65, 67, 68, 89, 105, 112, 115, 130
DNA polymerase, 35, 42
DNA repair, 49
DNA sequencing, 21, 103
DNAs, v, 2, 7, 16, 29, 33, 34, 35, 36, 46, 48, 49, 52, 54, 55, 56, 57, 69, 71, 79, 100, 101, 102, 109, 110, 112, 113, 114, 116, 119, 128, 129, 130, 132, 133, 138, 147
doctors, 105
dogs, 21, 149
dominance, 137
DOP, 37, 38
dosage, 134
double helix, 2
draft, 67, 73, 87, 105, 120, 126
Drosophila, 1, 7, 23, 26, 37, 54, 56, 61, 65, 66, 73, 100, 103, 110, 111, 112, 114, 115, 116, 117, 119, 123, 124, 126, 127, 128, 129, 130, 131, 133, 138, 139, 141, 146, 147, 148, 151
drugs, 105
dyes, 3, 16
dynamism, 22, 28

E

egg, 1, 15
electron, 4
electron microscopy, 4
elephants, 82
elongation, 71, 74
elucidation, 19, 31, 45, 63, 74, 88
embryogenesis, 127
emission, 46, 47
encoding, 25, 49, 50
endonuclease, 42, 115
England, 33, 39
environment, 27, 89
environmental change, 28
environmental factors, 90
environmental influences, 117
enzyme, 7, 35, 37, 42
enzymes, 26, 36, 42
epidemiology, 56
equipment, 32
erythrocytes, 131
EST, 41
euchromatin, 3, 4, 18, 65, 119, 126
eukaryote, 49, 56
eukaryotic, 2, 5, 21, 29, 49, 63, 64, 65, 68, 111, 119, 126, 140, 148
eukaryotic cell, 2, 5

evidence, 1, 27, 59, 62, 63, 67, 68, 73, 94, 97, 111, 113, 116, 124, 129, 132, 134, 137, 138, 149, 150
excision, 69
excitation, 46, 47
exons, 24, 26, 27
exonuclease, 42
extraction, 36, 44

F

families, 10, 11, 12, 14, 24, 26, 34, 49, 52, 53, 54, 55, 56, 57, 58, 59, 60, 63, 67, 68, 82, 84, 88, 100, 110, 112, 113, 114, 117, 125, 134, 138, 143, 144, 147, 149
fertilization, 98, 147
fiber, 4, 44, 46, 59, 71, 75, 76, 77, 144, 145, 148
fibers, 3, 4, 6, 34, 46, 76, 77, 115
filters, 47
fish, 8, 10, 11, 15, 16, 17, 18, 19, 22, 27, 33, 36, 37, 39, 45, 53, 55, 58, 61, 66, 67, 68, 73, 89, 92, 93, 95, 96, 97, 98, 99, 100, 106, 110, 111, 116, 117, 125, 130, 131, 132, 134, 135, 136, 137, 138, 141, 143, 144, 146, 151
fission, 13, 25, 26, 86
fitness, 23
fixation, 92
flatworms, 22
fluorescence, 32, 34, 38, 46, 47, 52, 54, 71, 74, 77, 113, 115, 123, 124, 125, 134, 137, 138, 141, 142, 143, 147, 148
fluorophores, 46, 47
force, 25, 112
formation, 25, 60, 95, 100, 145
formula, 46, 131
foxglove, 41
fragility, 140
fragments, 25, 40, 43, 61, 99, 132
freshwater, 11, 110, 116, 118, 135
fungi, 67, 90, 98, 120, 122
fungus, 6
fusion, 1, 11, 25, 26, 63, 82, 86, 87, 97, 150

G

gamete, 98
GC-content, 21
gel, 36
gene expression, 6, 23, 26, 103, 121, 122
gene mapping, 92, 93, 120, 133, 143
gene regulation, 49
gene transfer, 26, 115, 124, 127
genes, 1, 3, 16, 18, 21, 24, 25, 26, 27, 29, 30, 34, 35, 36, 41, 49, 51, 52, 54, 55, 61, 66, 69, 71,

72, 73, 74, 76, 77, 80, 89, 90, 92, 93, 94, 95, 99, 100, 102, 105, 109, 110, 111, 113, 115, 116, 117, 118, 121, 122, 123, 125, 127, 128, 129, 130, 131, 133, 135, 138, 139, 140, 141, 142, 143, 145, 146, 150
genetic defect, 74
genetic diversity, 120
genetic information, 1, 21, 42
genetic linkage, 73, 121
genetic marker, 56, 115
genetic traits, 1
genetics, iv, 1, 43, 56, 62, 72, 73, 137
genomic regions, 3, 69, 76
genomics, 29, 40, 44, 73, 77, 81, 102, 105, 106, 111, 122, 126, 133, 134, 138, 140
genotyping, 133
genus, 10, 15, 22, 35, 57, 58, 59, 60, 61, 63, 69, 96, 97, 101, 110, 111, 113, 115, 122, 124, 127, 139, 141, 142
germ cells, 1
germ line, 25
giardia, 72
gonadal dysgenesis, 74
gonads, 89
guanine, 31

H

haploid, 14, 22, 29, 60, 90, 98
harbors, 96
health, 105, 111
Helicobacter pylori, 56
heredity, 1, 133
heterochromatin, 3, 4, 16, 19, 22, 23, 35, 37, 55, 57, 58, 59, 60, 61, 65, 66, 69, 75, 76, 80, 88, 96, 99, 111, 113, 114, 115, 116, 122, 124, 125, 126, 131, 134, 137, 138, 144, 149
heterochromatinization, 99
heterogeneity, 57
histone, 3, 4, 33, 35, 36, 49, 52, 54, 55, 100, 109, 113, 114, 117, 118, 122, 123, 127, 131, 132, 133, 137, 138, 140, 141, 142, 145, 146, 151
histones, 4, 6, 54, 55, 77
history, 1, 2, 10, 30, 69, 83, 87, 112, 146
homologous chromosomes, 91, 144
hormone, 90
horses, 21, 82
host, 26, 66, 67, 69, 98, 99, 115
hot spots, 120
hotspots, 60
human, 2, 13, 24, 29, 30, 39, 44, 45, 46, 54, 56, 61, 62, 63, 65, 67, 68, 72, 73, 75, 76, 77, 79, 80, 81, 82, 83, 84, 85, 92, 105, 106, 109, 110, 111, 112, 113, 115, 116, 118, 120, 121, 125, 128, 129, 134, 137, 138, 141, 142, 145, 148, 149, 150
human genome, 24, 62, 63, 80, 105, 106, 116, 125, 128, 129
hybrid, 44, 101, 120, 121, 141
hybrid cell, 44
hybridization, 2, 31, 41, 43, 44, 46, 53, 61, 68, 70, 72, 73, 74, 75, 77, 79, 92, 101, 115, 117, 119, 126, 129, 133, 136, 138, 140, 143, 144, 145, 146, 148, 149
hydrogen, 31
hydrogen bonds, 31
hypothesis, 10, 13, 20, 59, 82, 86, 87, 94, 96, 100, 144

I

ideal, 23
identification, 10, 16, 19, 27, 30, 34, 36, 43, 52, 58, 73, 75, 79, 80, 90, 105, 117, 121, 126, 129, 137
illumination, 47
image, 47
images, 47
immune system, 25
improvements, 45
in situ hybridization, 2, 16, 20, 31, 32, 33, 34, 40, 41, 44, 52, 54, 71, 74, 77, 100, 110, 113, 123, 124, 125, 126, 131, 134, 136, 137, 138, 139, 141, 142, 143, 144, 147, 148, 149, 150
in vitro, 7, 42, 112, 118, 120
incidence, 26
incompatibility, 22
individuality, 1
individuals, 15, 66, 99
inferences, 24, 106
inheritance, 1, 28, 124, 127, 133
initiation, 56
insects, 9, 10, 14, 15, 16, 22, 25, 35, 36, 37, 40, 44, 52, 53, 55, 57, 62, 65, 67, 89, 90, 113, 125, 136, 143
insertion, 25, 54, 69, 91, 97, 117, 132, 140
integration, 2, 29, 30, 31, 40, 56, 72, 103, 106, 150
interference, 41
interphase, 3, 4, 6, 16, 59, 72, 75, 76, 125, 129
intervening sequence, 134
inversion, 19, 95, 97
invertebrates, 9, 10, 14, 15, 17, 18, 24, 27, 34, 52, 53, 55, 61, 66, 79, 80
iris, 119
Islam, 151
isochromosome, 58, 100, 132

isolation, 35, 36, 37, 39, 44, 53, 103, 123, 142, 151
Italy, 114

K

karyotype, 8, 10, 12, 13, 14, 16, 22, 24, 30, 38, 39, 45, 59, 60, 61, 63, 80, 81, 82, 83, 84, 85, 86, 87, 88, 93, 94, 118, 125, 128, 129, 130, 133, 139, 140, 144, 150
karyotyping, 31, 45, 113
kinetochore, 6, 146
kinship, 56, 111

L

labeling, 31, 35, 41, 42, 43, 44, 46, 71, 74, 79, 119, 120, 127, 136, 137, 145, 148
landscape, 68, 148
larvae, 11
lead, 23, 28, 89, 100, 101
lens, 47
Lepidoptera, 134, 147
leukemia, 105
light, 15, 18, 47, 77, 90, 93
liver, 23
localization, 59, 61, 66, 71, 73, 74, 110, 121, 124, 130, 131, 136, 141, 142, 143, 151
loci, 52, 54, 61, 65, 74, 90, 113, 117, 121, 123, 124, 128, 130, 136, 141, 147
locus, 34, 40, 56, 66, 71, 73, 76, 77, 96, 144, 145
lying, 24

M

machinery, 6, 7
magnitude, 62
major histocompatibility complex, 77, 111
majority, 12, 22, 26, 60, 61
malignant melanoma, 123
mammal, 8, 13, 17, 19, 23, 24, 39, 45, 58, 60, 69, 80, 84, 86, 92, 93, 100, 101, 116, 120, 123, 147
mammals, 10, 13, 16, 18, 19, 24, 35, 36, 37, 39, 40, 52, 56, 60, 61, 62, 67, 68, 79, 80, 81, 84, 86, 87, 88, 89, 90, 92, 94, 98, 106, 113, 122, 131, 134, 135, 140, 142, 144, 147, 148, 150
man, 74, 145
manipulation, 2, 35, 39
mapping, 16, 27, 29, 32, 35, 41, 46, 55, 56, 57, 58, 59, 61, 67, 68, 69, 71, 72, 73, 74, 75, 76, 77, 79, 80, 81, 82, 88, 94, 95, 96, 97, 99, 100, 103, 106, 109, 113, 114, 115, 116, 120, 122, 123, 124, 125, 127, 128, 129, 132, 137, 141, 145, 146, 147, 148, 150

marine environment, 11
marine fish, 11, 120, 133
matrix, 4, 6, 145
matter, iv
Mediterranean, 149
meiosis, 5, 14, 23, 24, 73, 92, 93, 102, 103, 144
messengers, 127
metaphase, 4, 44, 45, 46, 68, 71, 72, 73, 75, 76, 129, 131
methodology, 18, 19, 36, 37, 38, 43, 44, 45, 71, 74, 75, 80, 87
methylation, 3, 4, 23, 54
Mg^{2+}, 42
MHC, 77, 124
mice, 7, 15, 23, 60, 76, 84
microsatellites, 56, 57, 61, 62, 111, 113, 118, 119, 136
microscope, 8, 35, 38, 41, 46, 47, 74, 107
microscopy, 29, 71, 74, 90
Middle East, 125
minisatellites, 55, 57, 61, 66, 119
Missouri, 112
mitochondria, 21
mitosis, 1, 3, 5, 23
models, 5, 21, 30, 66, 69, 84, 93, 111, 116, 120, 125
modifications, 4, 6, 31, 37, 39, 43, 54, 71, 128
molds, 119
mole, 82, 140
molecular biology, 1, 2, 3, 29, 32, 34, 102, 105, 143
molecular cytogenetics, 2, 30, 31, 79, 92, 94, 102, 112, 120, 122, 149
molecular dating, 121
molecular dynamics, 140
molecular medicine, 132
molecules, 19, 25, 32, 41, 106, 120
mollusks, 9, 36, 40, 52, 53, 55, 58
monomers, 57, 60
Morocco, 101
morphology, 2, 8, 9, 10, 11, 12, 13, 15, 31, 38, 76, 105, 139, 144
motif, 40, 57, 61, 62, 117, 142
mutant, 130
mutation, 92, 119, 129
mutations, 22, 25, 27, 99
myoglobin, 56, 148

N

nanometer, 4
natural selection, 66
negative consequences, 69
nematode, 22, 23, 67, 130, 134, 145

neutral, 61
New Zealand, 15, 122, 142
next generation, 2, 25, 57, 105, 106
Nile, 58, 72, 73, 76, 96, 114, 119, 123, 126, 128, 129, 135
nuclear genome, 21, 111
nuclei, 1, 75, 76, 145
nucleic acid, 71, 134, 148
nucleolus, 147
nucleosome, 4, 54
nucleotide sequence, 2, 29, 30, 36, 55, 59, 60, 88, 105, 106, 117
nucleotide sequencing, 35
nucleotides, 19, 40, 41, 42, 55, 61, 74
nucleus, 3, 21, 22, 34, 43, 59
null, 130

O

oogenesis, 54
opportunities, 22, 66
organelles, 26
organism, 21, 24, 27, 44, 49, 57, 90, 105
organize, 6
osteology, 130
oyster, 58, 61, 116, 133, 148
oysters, 123

P

pachytene, 46, 75, 76, 77, 115, 135
Pacific, 58, 113, 116, 123, 131, 134, 148
paints, 19, 39, 45, 48, 97, 122, 123, 139
pairing, 6, 56
paleontology, 126
parallel, 6
parallelism, 114
parasite, 98
parasites, 25
parthenogenesis, 149
pathology, 148
pathways, 14, 77, 89, 95, 97, 116
PCP, 79, 97
PCR, 32, 33, 34, 35, 36, 37, 38, 41, 42, 43, 67, 74, 132, 138, 144, 148
penalties, 145
peptides, 26, 127
pericentric inversion, 129
permit, 30, 41, 44
phenotype, 89, 90, 98, 99, 102
phosphate, 31
phosphorylation, 6, 54
phylogenetic tree, 122
phylum, 117

physical properties, 71
physical structure, 105
pigs, 21, 83, 138
plants, 7, 22, 26, 31, 44, 61, 67, 76, 90, 92, 98, 109, 120, 122, 125, 151
plasmid, 36
plasticity, 22, 25, 28
platform, 145
Platyhelminthes, 98
playing, 54
ploidy, 23, 49, 112
polymerase, 42, 52, 71, 74, 123
polymerase chain reaction, 74
polymerization, 35
polymorphism, 10, 12, 15, 16, 18, 110, 112, 119, 134
polymorphisms, 8, 25, 27
polyploid, 22, 23
polyploidy, 12, 22, 23, 27, 144
polythene, 37
population, 56, 57, 62, 95, 100, 103, 109, 135
positive correlation, 24
preparation, iv, 4, 29, 45, 74, 76
president, 105
prevention, 105
primate, 26, 57, 82, 83, 109, 116, 120, 135
priming, 42, 43
PRINS, 71, 74, 115, 116, 120, 121, 127, 136, 137, 145, 148, 149
probability, 1
probe, 31, 33, 34, 35, 36, 37, 38, 40, 41, 42, 43, 45, 46, 47, 70, 71, 73, 76, 80, 86, 88, 92, 93, 95, 96, 97, 98, 99, 100, 101, 117, 118
project, 115
prokaryotes, 119
promoter, 52
prophase, 75
proposition, 92
protection, 7
protein sequence, 36
protein synthesis, 52, 66
protein-protein interactions, 25
proteins, 3, 4, 5, 6, 7, 16, 21, 25, 26, 54, 65, 76, 110, 123, 127, 144, 147
proteome, 151
proto-oncogene, 121
purification, 38
pyrimidine, 41

Q

quinacrine, 19

R

radiation, 121
Radiation, 114
radioactive isotopes, 32
reading, 42
reagents, 76, 77
reciprocal cross, 45, 79
reciprocal translocation, 138
recognition, 4, 18, 19, 24, 38, 41, 57
recombinant DNA, 129
recombination, 1, 26, 29, 52, 56, 57, 68, 69, 73, 76, 90, 91, 96, 102, 103, 110, 112, 115, 117, 120, 128, 144, 147
recommendations, iv
reconstruction, 30, 125
redistribution, 57
redundancy, 22
regression, 119
relatives, 23
replication, 7, 19, 42, 49, 56, 110
reproduction, 22, 62, 89, 130
reproductive organs, 23
reptile, 24, 72, 127, 136, 141
researchers, 90, 98, 103
residues, 6, 7
resistance, 140
resolution, 31, 32, 43, 44, 46, 48, 71, 72, 75, 76, 86, 113, 120, 121, 123, 128, 132, 139, 141, 148
resources, 71, 109
response, 28
restriction enzyme, 35
restructuring, 60
retrovirus, 69
reverse transcriptase, 7, 149
ribosomal RNA, 49, 51, 52, 111, 113, 118, 131
risk, 105
RNA, v, 2, 7, 26, 35, 42, 49, 51, 52, 63, 106, 111, 117, 119, 120, 123, 127
RNAs, 2, 52
Robertsonian translocation, 14, 140
rodents, 13, 44, 68, 81, 87, 146
roots, 105
roundworms, 23
routes, 26
rules, 54

S

salivary gland, 15
salmon, 22, 55, 61, 72, 73, 117, 137, 143, 145
saltwater, 142
scarcity, 62, 65
science, 1, 105
scientific knowledge, 26
scientific papers, 29
scope, 81
secretion, 90
segregation, 1, 6, 14, 23, 49, 61, 99, 103, 135
self-fertilization, 22
sensitivity, 31, 71, 74, 77, 138, 145
sequencing, 21, 27, 36, 39, 40, 44, 52, 68, 69, 72, 77, 80, 100, 102, 106, 109, 115, 128, 131, 148
serine, 124
sex chromatin, 120
sex ratio, 98, 132, 146, 148
sex reversal, 90
shape, 7, 13, 20, 80
sheep, 21, 73, 77, 83, 117, 121
showing, 14, 17, 21, 27, 39, 44, 48, 59, 60, 73, 76, 80, 89, 92, 93, 95, 97, 100, 101, 111
sibling, 148
signals, 34, 35, 41, 42, 47, 53, 60, 61, 66, 68, 69, 71, 72, 73, 91, 92, 97
silkworm, 65, 144
silver, 16, 17, 18, 19, 31, 53
snakes, 12, 94, 95, 97, 131, 135, 142
SNP, 124
solid tumors, 126
solution, 16, 18, 42
somatic cell, 56
South America, 82, 96, 116, 139, 142
Southern blot, 36, 131
Spain, 101
speciation, 11, 23, 62, 139
specific tax, 96
speech, 105
sperm, 1, 15, 23
spermatogenesis, 135
spindle, 6
stability, 7, 56, 59, 68, 72, 87, 94, 149
stabilization, 7, 66
stars, 105
stasis, 11, 118, 133, 135
state, 5, 12, 47, 91
states, 3, 5
storage, 36
stretching, 76, 77
structural changes, 71
structural variation, 24, 44
structure, 1, 4, 6, 7, 8, 9, 10, 11, 12, 13, 14, 16, 20, 24, 25, 31, 35, 36, 39, 44, 52, 54, 55, 57, 60, 63, 65, 69, 74, 77, 103, 110, 130, 132, 138, 143, 146, 148, 151

substitution, 41
substrate, 124
Sun, 76, 128, 144
suppression, 76, 90, 91, 103, 129, 148
surface area, 23
surplus, 66
survival, 125
symmetry, 27, 39
syndrome, 74
synthesis, 7, 41, 42, 52, 74, 121, 127, 149

T

tandem repeats, 7, 27, 52, 57, 60, 119, 138, 147
target, 31, 35, 38, 41, 43, 71, 74, 75, 76, 102, 105, 144
taxa, 10, 22, 23, 24, 30, 35, 36, 59, 61, 97, 130
techniques, 2, 10, 12, 14, 16, 20, 21, 29, 31, 32, 34, 35, 71, 72, 77, 99, 102, 124, 126, 132, 137, 148
technologies, 3, 10, 20, 21, 71, 102, 103, 105, 107
technology, 34, 37, 39, 72, 79, 100, 132
telocentric chromosome, 12
telomere, 6, 7, 11, 25, 34, 35, 62, 65, 66, 75, 76, 77, 135, 140, 144, 149
temperature, 89, 94, 95
testing, 36, 56
testis, 89, 95, 111, 128, 129
tetrapod, 27
thymine, 31, 41
tissue, 2, 29, 112, 132
traits, 21, 57, 89, 90, 92, 94, 95, 96, 98, 99, 102
transcription, 3, 24, 25, 42, 56, 65, 118, 123
transcription factors, 24
transcripts, 52
transformation, 12, 28
translation, 36, 42, 43
transmission, 47, 67, 99
treatment, 16, 18, 37, 74, 105
triggers, 89
triploid, 22, 111, 112, 123
trisomy, 138
trypsin, 18
tumor, 44, 128
turnover, 67, 128, 133
turtle, 95, 118, 131

U

ultrastructure, 6
uniform, 12, 61, 65
United, 109
United States (USA), 33, 105, 109, 113, 115, 116, 117, 118, 119, 120, 122, 125, 126, 128, 129, 131, 136, 138, 139, 143, 146, 150

V

variations, 10, 11, 14, 36, 44, 54, 56, 57, 58, 74, 77, 133
vector, 127
vertebrate classes, 80
vertebrates, 7, 9, 10, 15, 17, 18, 21, 23, 24, 26, 27, 34, 40, 53, 58, 66, 67, 75, 76, 79, 87, 92, 95, 97, 99, 106, 111, 112, 122, 127, 137, 139, 142
visualization, 41, 70, 107

W

water, 66, 87, 101, 135, 138, 142
wavelengths, 46, 47
wealth, 52
web, 81
whales, 83
windows, 102
wood, 100
workers, 30
worms, 89
WWW, 140

X

X chromosome, 45, 57, 61, 68, 73, 74, 81, 92, 93, 94, 96, 97, 100, 101, 103, 123, 135

Y

Y chromosome, 61, 69, 89, 90, 92, 94, 96, 97, 103, 114, 122, 123, 143
YAC, 71, 72, 76, 123
yeast, 3, 7, 71, 142
yolk, 75

Z

zoogeography, 130